North American Mixing Forum 2018

Core Programming Area at the 2018 AIChE Annual Forum

Pittsburgh, Pennsylvania, USA
28 October - 2 November 2018

ISBN: 978-1-5108-7627-9

Printed from e-media with permission by:

Curran Associates, Inc.
57 Morehouse Lane
Red Hook, NY 12571

Printed by Curran Associates, Inc. (2019)

AIChE
120 Wall Street, FL 23
New York, NY 10005-4020

Phone: (800) 242-4363
Fax: (203) 775-5177

www.aiche.org

Additional copies of this publication are available from:

Curran Associates, Inc.
57 Morehouse Lane
Red Hook, NY 12571 USA
Phone: 845-758-0400
Fax: 845-758-2633
Email: curran@proceedings.com
Web: www.proceedings.com

TABLE OF CONTENTS

Some abstracts in this book refer to images that were not made available by the publisher

532790 Mixing Efficiency in a Simple, Continuous, Laminar-Flow Microdevice Using Computational and Experimental Approaches

Abstract:

Microfluidic devices are small-scale designed to mix two miscible fluids at the microscale level achieve in order to promote fluid mixing and blending by increasing the contact area between the mixing species. In a number of cases these devices use complex 2D or 3D architectures, including multiple splitting, recombining, or rotating channels incorporating structures with twists, embedded barriers, or staggered herringbone structures. These devices can be complex and expensive to fabricate. In this work, a simple but effective micromixer device was assembled to obtain a microfluidic mixer that is easy to fabricate, inexpensive, and readily built in most laboratories without the need for dedicated personnel, extensive training, and dedicated and costly fabrication equipment. To achieve this goal, serpentine micromixers were developed, built, experimentally tested, and computationally analyzed using a lamination-based assembly that can be obtained by using simple tools and supplies available in non-microdevice dedicated laboratories. A fluorescence imaging technique was then used to quantify the mixing effectiveness of these devices. Computational Fluid Dynamics (CFD) modeling of the same devices using a commercially available software (COMSOL) was conducted to both validate the computational approach and to corroborate the experimental results. The computational results were found to be in excellent agreement with the experimental data and showed that the serpentine micromixer can achieve significant levels of mixing efficiency. In addition, different modified versions of the same basic serpentine design were analyzed using CFD in order to determine which systems would result in improved mixing efficiency. Finally, computational results for a straight microfluidic channel were separately obtained in order to quantify the increase in mixing efficiency produced by the serpentine design with respect to a similar channel of the same length. These computational results were also compared with previous theoretical predictions for straight channels and were in excellent agreement with them.

536991 Effect of Shear Gap Width on Flow and Power Draw in an Inline Rotor-Stator Mixer

Abstract:

Rotor-stator mixers are close tolerance devices with a narrow â€œshear gap widthâ€ or clearance between the rotor tip and the inner stator surface. It is often recommended that they be scaled on nominal shear rate (ratio of rotor tip speed to shear gap width) but this is often inadequate. Pumping rate is a factor and in turbulent flow energy dissipation rates are highest in the mixing layers and strong stator slot jets. These are created when fluid leaving the rotor abruptly changes direction as it impinges on the leading edge of the stator slot and is forced through it. The effect of shear gap width on rotor-stator process performance is not well studied.

Three-dimensional computational fluid dynamics (CFD) simulation was performed for water in turbulent flow to investigate the effect of shear gap width on the flow field, pumping rate and power draw of a Silverson L4R bench scale inline rotor-stator mixer equipped with a square hole stator head. Sliding mesh simulations of the Reynolds-Averaged Navier-Stokes (RANS) equations were performed using Fluent with the realizable k-Îµ turbulence model and enhanced wall functions. The power draw was determined from the rotor torque and the pumping capacity was calculated from the velocity exiting the stator slots.

The stator dimensions were held constant (inner diameter = 31.3 mm). Several shear gap widths were investigated ranging from the standard shear gap width w = 0.2 mm to the widest one with w = 1.7 mm. It was found that increasing the shear gap width caused a small decrease in pumping capacity. However, the mill head with the standard shear gap draws the most power and power draw decreases significantly with increasing shear gap width. Flow field features such as mean velocity, turbulent kinetic energy and energy dissipation rate are significantly influenced by shear gap width, particularly the extent to which the stator jets penetrate the volute. Examining these quantities provides physical insight into the flow and power results. Motivation for the study will be presented and use of the results will be illustrated by application to the analysis of crystal wet milling experiments performed in these devices.

Comparison of Breakup Kernels in the CFD-PBM Simulation of a Pulsed Disc and Doughnut Column

Yu Xiong, Li Shaowei, Jing Shan, Tsinghua University, Beijing, PRC

Introduction

Liquid-Liquid flow is a commonly found chemical engineering process where one liquid phase is dispersed as droplets in the other phase to enhance mixing Solvent extraction, as a typical example of liquid-liquid flow, has been used in hydrometallurgy, environmental engineering and nuclear fuel reprocessing fields. Pulsed Disc and Doughnut Column (PDDC) is a recently developed liquid-liquid extraction equipment to recovery valuable materials. For its high throughput, no rotating parts and self-cleaning properties, PDDC promises excellent operation characteristics.

Determination of droplet size distribution (DSD) is of good significance for the research and design of PDDC. Population balance model (PBM) is the most promising method to predict the DSD of liquid-liquid system. Coupled with computational fluid dynamics (CFD), the evolution of droplet diameter is directly derived from the solution of population balance equations. CFD simulation provides the general parameters of flow field while PBM offers a variable droplet size to calculate phase momentum and mass interactions. It gives a refined image of the complex liquid-liquid flow in extraction equipment's. CFD-PBM method processed well in recent years, but its applications on PDDC are rarely reported due to the lack of proper kernel functions to describe droplet behaviors.

Application of CFD-PBM method to predict droplet size distribution depends on the kernel functions to describe breakage and coalescence events of droplets. Many droplet breakup and droplet-droplet coalescence functions have been derived theoretically or correlated experimentally. Although kernel functions proposed by these researchers are frequently used, their feasibility to the CFD-PBM simulation of two-fluid flow in PDDC needs further investigation.

In view of the lack of suitable PBM kernel functions in PDDC, we proposed an experimental correlation for droplet breakage kernel functions in PDDC. This is the first PBM kernel functions particularly designed for the PDDC, precisely, based on directly measured droplet breakage data The main purpose of this work is to couple these droplet breakage kernel functions into the CFD-PBM simulation of PDDC and investigate the liquid-liquid flow behaviors.

Experiments

Experiments were carried out in a square-sectioned pulsed disc and doughnut column with glass windows on both sides to allow better visual observations. Figure 1a shows the diagrammatic sketch of the experimental set-up.

Pulsation intensity, viscosity of dispersed phase and interfacial tension are the key factors affecting the droplet breakage behaviors. Distilled water with various amount of glycerol were set as dispersed phase to investigate the influence of viscosity. Kerosene with different volume percentage of TBP as continuous phase to study the effects of interfacial tension.

Firstly, the column was filled with the continuous phase. Subsequently, the dispersed phase was introduced from the top of the column by a diaphragm pump while the pulsation generator was working. The dispersed fluid then was broken into droplets by the inner circulation of the continuous fluid and coalesced at the bottom of the column. The periodically steady state could be reached after enough time at the given flow rate. After that, the droplet behavior was observed by a micro-scope connected to a high speed camera, which took the photographs using a frame rate of 1000 frames per second. The smallest drop size that can be observed by the high speed camera is approximately 0.1mm. Drop breakage was focused at three different positions of the column. The distances between the observed sections and the central section of the column were 0 mm, 20 mm and 40 mm, respectively, as shown in Figure 1b.

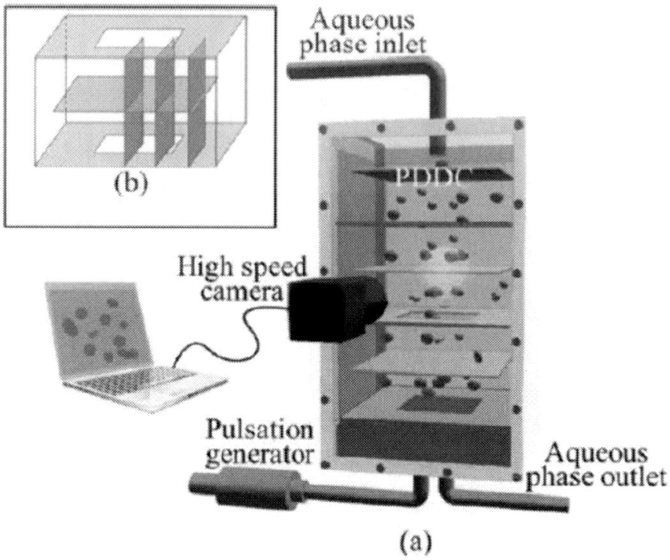

Figure 1. Schematic diagram of experiment setup. (a) Pulsed disc and doughnut columns; (b) observation zones.

The droplet size was measured manually and breakup events were counted artificially. As small flowrate of dispersed phase was used, the droplet-droplet coalescence event was neglected. At the observed volume V and the recording period T, the number of droplets within size range $d+\Delta d$ gives the droplet size distribution function $n(d)$. The number of droplets breakup within size range $d+\Delta d$ derives the droplet breakage frequency function $b(d)$ and the diameter of the mother droplet d_m and daughter droplet d daughter size distribution function $\beta(d|d_m)$.

Simulation

Comparison of breakup kernel functions

Kernel functions concerned to droplet breakage events are droplet breakage frequency function $b(d)$, daughter size distribution function $\beta(d|d_m)$. Two PBM kernel functions, one is derived by Coulaloglou& Tavlarides, which is widely used by numerous researchers, the other is the correlations via experiments in PDDC from our experiment, are embedded in the CFD code. The expressions of these functions are listed in Table 1. Two adjustable parameters and are found in the model of C&T. It is recommended to be 0.15 and 1.29, respectively.

Before the implementation of the PBM into CFD, a brief illustration and comparision about these two models is conducted.

Table 1. Droplet breakup frequency and daughter droplet size distribution functions of C&T and our correlations.

Researcher	Mathematic expressions
Coulaloglou& Tavlarides (1977)	$b(d)=c_1\frac{\varepsilon^{1/3}}{(1+\phi)d^{2/3}}\exp\left(-c_2\frac{\gamma(1+\phi)^2}{\rho_d\varepsilon^{2/3}d^{5/3}}\right)$
	$\beta(d\mid d_m)=-7.2\frac{d^2}{d_m^3}\exp\left(-4.5\frac{\left(2d^3-d_m^3\right)^2}{d_m^6}\right)$
Our correlations	$b(d)=0.0345\frac{Af}{h}\left(Af\left(\frac{\rho_c}{g\gamma}\right)^{0.25}\right)^{2.45}\left(Af\frac{\rho_c}{\mu_d}d\right)^{2.9}$
	$\beta(d\vert d_\mathrm{m})=\begin{cases}14.55\frac{d^2}{d_\mathrm{m}^3}\left(\frac{d^3}{d_\mathrm{m}^3}\right)^{0.59}\left(d_\mathrm{m}\left(\frac{\rho_c g}{\gamma}\right)^{0.5}\right)^{-0.049}, & \frac{d^3}{d_\mathrm{m}^3}\leq 0.5 \text{ and } d_\mathrm{m}\\ 2.88\frac{d^2}{d_\mathrm{m}^3}\left(\frac{d^3}{d_\mathrm{m}^3}\right)^{-0.40}\left(d_\mathrm{m}\left(\frac{\rho_c g}{\gamma}\right)^{0.5}\right)^{0.17}, & \frac{d^3}{d_\mathrm{m}^3}\leq 0.5 \text{ and } d_\mathrm{m}\end{cases}$
	$d_\mathrm{mc}=1.22\frac{\mu_d}{Af\rho_c}\left(Af\left(\frac{\rho_c}{g\gamma}\right)^{0.5}\right)^{-0.89}$

One can note from Figure 2 that both models tell breakage frequency increases with larger droplet diameter d and turbulent energy dissipation ε. With proper parameters, the breakage frequency values are similar predicted by these two models. However, as droplet diameter increase, the model of C&T predicts the breakage frequency to go up slowly, while our model tells that droplet size affects the breakage frequency strongly. Droplet breakage is an energy demanding process. Interfacial energy and turbulent energy dissipation provide the main energy source for droplet breakage in liquid-liquid flow. Larger droplets reserve higher interfacial energy, inducing surface instability to promote droplet breakage. Turbulent energy dissipation transports the kinetic energy of main fluid to the droplets.

For the daughter droplet size distribution function, our model is an empirical correlation. In this model, a critical droplet diameter d_{mc} is defined. It is assumed that droplets smaller than this size tend to breakup unequally with a U-shape distribution and prefer to split equally leading to an inverted U-shape distribution if larger than d_{mc}.

The daughter droplet size distribution is a truncated normal function in C&T's model. It is originally a symmetrical function regarding to the volume ratio (defended as daughter droplet volume to mother droplet volume). A transformation is performed to match up the with droplet breakage frequency function with droplet diameter *d* as argument. In C&T's model, it only concerns to the size of mother and daughter droplet and is independent to the local hydrodynamic conditions. Similar distribution curves are then expected for mother droplet with different diameters, as shown in Figure 2.

The critical droplet diameter is 2.37 mm when turbulent energy dissipation rate ε is 0.107 m^2/s^3. Three typical droplet diameter sizes, that is, 1.6 mm, 3.0 mm, and 5.4 mm are chosen to evaluate the differences between these two models. As mentioned, according to C&T's model, they present the identical features: a maximum value at diameter of $0.5^{1/3}$ d_m and nearly negligible distributions when daughter droplet size is less than 0.5 *d*. A narrow droplet size distribution is expected with this distribution pattern. Since mother droplets with size 1.6 mm is smaller than the critical droplet diameter, it tends to break unequally according to our model, this leads to the distribution of monotone increasing function as shown Figure 2. For mother droplets with size 3.0 mm or 5.4 mm, a maximum value is promised at $0.5^{1/3}$ d_m. the same at C&T's model. This originates from the assumption that mother droplets are more likely to breakup equally under this condition. Moreover, the probability of distributions to small size droplets is much greater than that at C&T's model. Differences between these two models might have a significant influence to the final distribution result.

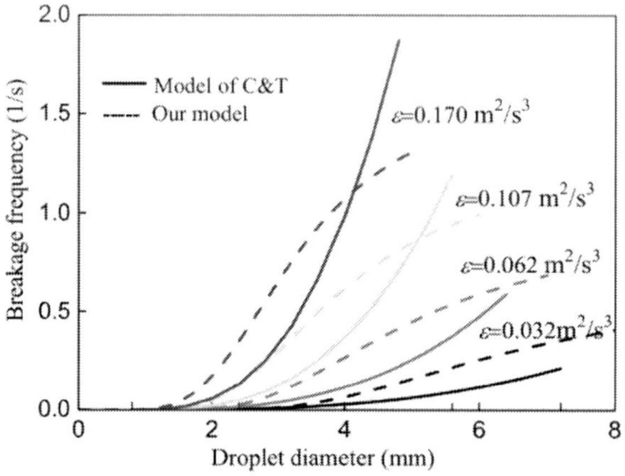

Figure 1. Comparison of the droplet breakup frequency function with various turbulent energy dissipation ε.

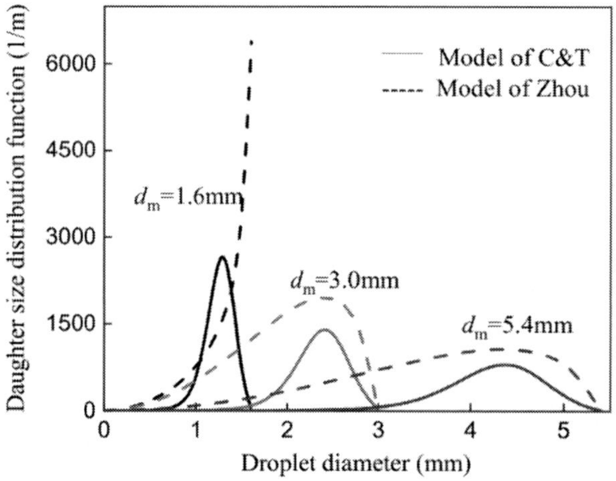

Figure 2. Comparison of the daughter size distribution function with various d_m.

Coupling of CFD and PBM

The coupling of PBM with CFD is based on the Euler-Euler framework. A local droplet diameter is required to solve the interfacial momentum transfer terms of governing equations. As only one set of N-S equations is solved, it means that the all assumed droplets with different pivot diameter share the same velocity field evaluated with the Sauter mean diameter d_{32}.

$$d_{32}(t)=\frac{\sum N_i(t)d_{p,i}^3}{\sum N_i(t)d_{p,i}^2}$$

In the description of CFD-PBM method, the droplet diameter parameter is not a global constant, conversely, it depends on time and locations. Precisely, if droplets of different size belong to different velocities, it is the complex case of multi-fluid flow. Under this consideration, the computational load is unborable to solve numerous governing equations with massive interactions between different phases. The CFD-PBM could be regarded as a compromise to balance the numerical accuracy of Euler-Euler method and computational cost of multi-phase flow.

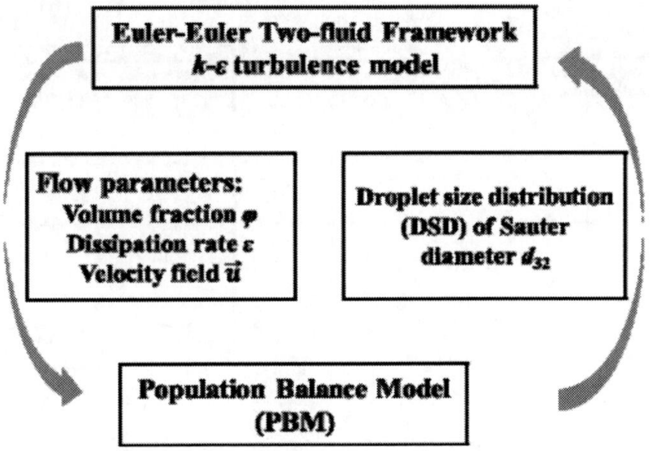

Figure 4. Coupling of CFD and PBM

Furthermore, the governing parameters to solve the population balance equations, such as turbulent energy dissipation rate ε, volume fraction α are the returned values of CFD results as shown in Figure 4. The droplet diameter to calculate mass and momentum transfer terms of governing equations is the derived results of PBM. Dependency d_{32} on correlations with physical or geometry parameters is eliminated. It is of great advantage to conduct the coupling CFD-PBM method on liquid-liquid flow simulations.

Results and Discussion

Cumulative droplet volume fraction

The simulation results by the two sets of PBM kernel functions are compared with the experimental results regarding to cumulative volume droplet size distribution Q_3.

One can see from Figure 5 that under the four operation conditions, results by our model fit more accurately with the experimental ones than that of C&T model.Experimental results indicate that middle size droplets contribute most percentage of the dispersed phase volume. The amount of larger droplets is fewer while the volume of smaller one is negligible, these lead to the "*J*" shape curve of the cumulative volume droplet size distribution, where two plateaus are distinct at the range of small and large droplet diameter, respectively. For the model of C&T, plateau at large droplet disappeared. It seems that C&T model underestimates the droplet breakage frequency, especially for larger size droplets as illustrated in Figure 2. At steady state, there are still great amount of large size droplets, which results in the deviation to experimental results

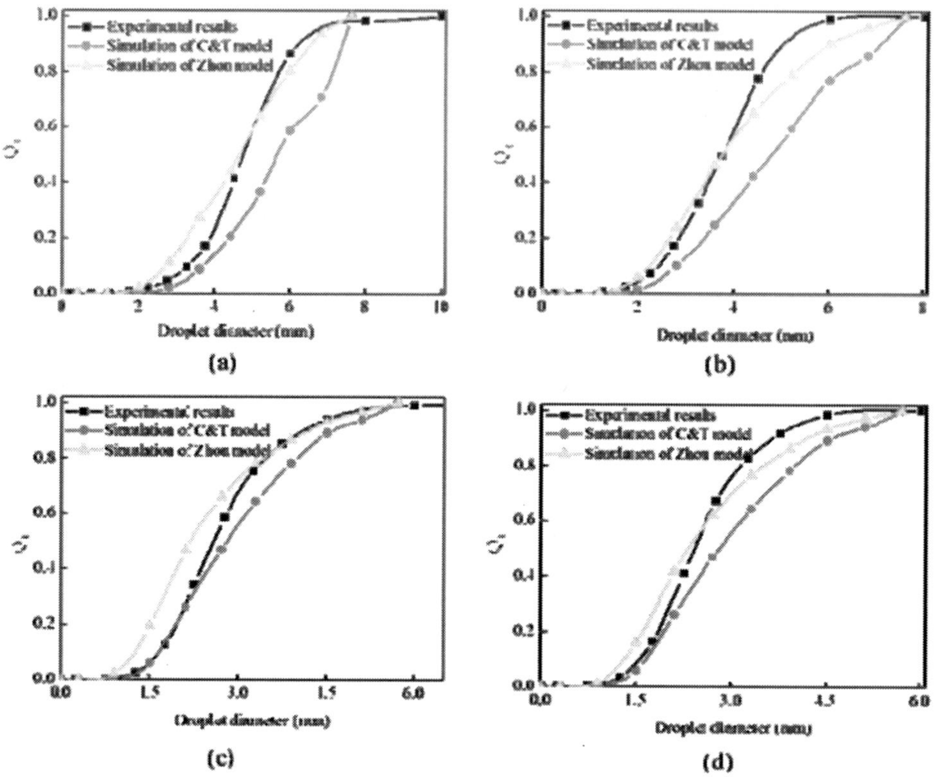

Figure 5. Comparison of experimental and simulation results on the steady cumulative volume droplet size distribution Q_3 at various operation conditions (pulsation intensity). (a) A=8mm, f=1.0Hz; (b) A=10mm, f=1.0Hz; (c) A=12mm, f=1.0Hz; (a) A=10mm, f=1.2Hz.

For our model, generally good agreements between experimental and simulation results are promised, this model captures the significant details of the droplet size distribution. As shown in Figure 2,our model predicts high probability of distributions to small droplets, this is also manifested in Figure 5, as simulation results of small size droplets by our model are always higher than that of experiments. This also causes the amount of large size droplets is fewer than experimental results.

It can be concluded from Figure 5 that PBM model of our correlations is more appropriate for the CFD-PBM simulations in PDDC. Since the model of C&T is derived semi-theoretically and contains two adjustable constants, it finds wider applications at the cost of accuracy in particular situations. In this view, simulation results by the model of Zhou is adopted in later investigations.

Hydrodynamic performance

The hydraulic performance of PDDC here is mainly the dispersed phase holdup and droplet mean diameter d_{32}. Both of these two parameters can be calculated from the droplet size distribution function, the experimental and simulation results are plotted in Figure 6 and Figure 7.

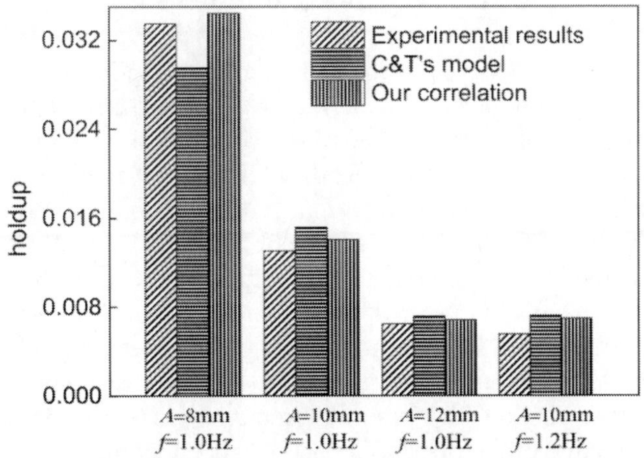

Figure 6. Comparison of experimental and simulation results on the dispersed phase holdup at various operation conditions (pulsation intensity).

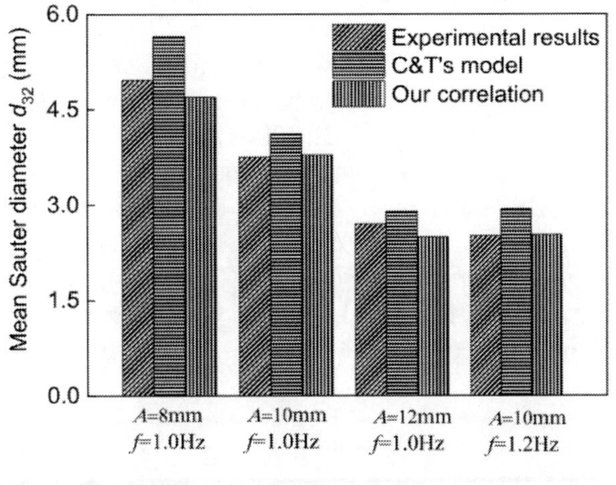

Figure 7. Comparison of experimental and simulation results on the mean Sauter diameter at various operation conditions (pulsation intensity).

As inlet flowrate of dispersed phase is not identical at the four cases, it is unable to analyze the influence of pulsation intensity on holdup. For droplet mean Sauter diameter, both experimental and simulation results predict that it decreases with higher pulsation intensity. This is in accordance with previous experimental work. Generally, just like that in Figure 5, our model predicts more accurately regarding to hydraulic performance. This corroborates that our PBM model is more appropriate to the CFD-PBM simulation in PDDC than that of C&T model. However, deviations between the results of two PBM models is not large enough to say which is suitable if we only investigate the dispersed phase holdup and droplet mean Sauter diameter. In other words, it is insufficient to determine whether one PBM model is suitable or not via the holdup and mean diameter results. Information directly read from the droplet size distribution can tell more details about the PBM model.

Summary and Conclusions

Via experiment work, we proposed a breakup kernel function in PDDC. Comparison with that of Coulaloglou & Tavlarides were conducted to describe droplet breakage behaviors. The differences are also directly verified by the simulation results regarding to cumulative volume droplet size distribution. Favorable agreements are obtained compared to the droplet size distribution predicted by our model and that measured by experiments. It suggests that our model is more appropriate to the PBM-CFD simulation for PDDC than that of C&T. However, the holdup and mean Sauter diameter by these two models do not show significant discrepancy with experimental results. It is thus suggested comparison of holdup and mean diameter is incapable to show the characteristics of different PBM models. Otherwise, the only purpose to introduce the PBM model is to provide the mean diameter to calculate drag force term. This is not the original idea to put massive calculation resources to coupling the PBM with CFD method. The important essence of CFD-PBM simulation method is to give deeper insights to droplet size distribution, for better knowledge of the operation performance the pulsed columns.

References

1. Coulaloglou, C.A., Tavlarides, L.L(1977), Description of interaction processes in agitated liquid-liquid dispersions, Chemical Engineering Science 32, 1289-1297.
2. Liu, J.-Q., Li, S.-W., Jing, S (2015). Hydraulic performance of an annular pulsed disc-and-doughnut column. Solvent Extraction and Ion Exchange 33, 385-406.
3. van Delden, M.L., Vos, G.S., Kuipers, N.J (2006a), Extraction of caprolactam with toluene in a pulsed disc and doughnut column part I: Recommendation of a model for hydraulic characteristics. Solvent. Extr. Ion. Exc. 24, 499-517.
4. Zhou, H., Jing, S., Fang, Q., Li, S., Lan, W J. (2017), Direct measurement of droplet breakage in a pulsed disc and doughnut column. AIChE Journal 63, 4188-4200.
5. Drumm, C., Attarakih, M.M., Bart, H.J (2009), Coupling of CFD with DPBM for an RDC extractor, Chemical Engineering Science 64, 721-732.

528123 Performance Characterization of the GS-4 Gas Induction Impeller

Abstract:

The GS-4 impeller is a novel gas induction impeller that is differentiated by its large openings for gas flow and generation of an axial flow pattern. This impeller is characterized based on the power number, pressure coefficient, and modeling of the induced gas flow rate. The power number at ungassed conditions is independent of impeller size and submergence, but is higher when up-pumping than down-pumping. This parameter decreases as gas is induced, and is modeled using the relative power number as a function of the relative speed. The pressure coefficient is examined using two approaches, one using speeds below the onset of induction and the other using the critical speed. Comparison of these methods shows that the pressure coefficient is more accurately determined independently of the minimum induction speed, and that there is a pressure loss encountered within the impeller as gas is forced outward from the center. This information is used to relate the available pressure difference to the gas flow rate. This relationship is affected by impeller diameter, with larger impellers inducing a higher flow rate than smaller impellers for a given pressure difference. The accuracy of the model is improved by assuming that the gas flow rate is also a function of the gas-liquid contact area within the impeller. The gas flow rate is also analyzed using a model that relates induction to an injection process, however it is difficult to evaluate the validity of this model due to the pressure drop introduced by the gas flow rate measurement method.

533862 The Effect of Scale-up on Mixing Efficiency in Oscillatory Flow Reactors Using Principal Component Based Analysis As a Novel Residence Time Distribution Measurement Tool

Abstract:

The effect of scale-up on mixing efficiency in oscillatory flow reactors using principal component based analysis as a novel residence time distribution measurement tool

Oscillatory flow strategies through baffled tubular reactors provide an efficient approach in improving process kinetics through enhanced micromixing and heat transfer[1]. Known to have high surface area to volume ratios, oscillatory flow baffled reactors (OFBR) generate turbulence by superimposing piston driven oscillatory flow onto the net flow generated by a pump. By tuning the oscillating parameters (amplitude and frequency), one can tailor the residence time distribution of the system for a variety of multiphase applications[2,3,4]. Using a microscope camera, principal component image analysis, and pulse tracer injections, a novel noninvasive approach has been developed to experimentally estimate dispersion coefficients in two geometrically different systems (DN6 and DN15, Alconbury Weston Ltd.). Similarly, a comprehensive experimental investigation of the effect of oscillation parameters on the residence time distributions (RTD) is discussed in both systems. The oscillation amplitude was found to have a significant positive correlation with the dispersion coefficient with 1 mm providing the least amount of dispersion in either system. Oscillation frequency had a less significant impact on the dispersion coefficient, but optimal operation was found to occur at 1.5 Hz for the DN6 and 1.0Hz for the DN15. Until now, OFBR literature has not distinguished between piston and pump driven flow. Pump driven flow was found to be ideal for both systems as it minimizes the measured dispersion coefficient. However, piston driven turbulence is essential for avoiding particle settling in two phase (solid-liquid) systems and should be considered in two phase applications such as crystallization.

To further understand how operating parameters affect system flow conditions, a computational fluid dynamics model was developed. Found to have good agreement with experimental validation, the CFD model further explains how system geometry and oscillation conditions drastically affect the size and shape of eddies generated, ultimately impacting the measured dispersion coefficient. The CFD model also sheds light on how the RTD of the system is (indirectly) dependent on temperature and mass diffusivity. Using a combined experimental and computational approach, optimal operating conditions were determined for each system. References

1. Stonestreet, P., & Harvey, A. P. (2002). A Mixing-Based Design Methodology for Continuous Oscillatory Flow Reactors. *Chemical Engineering Research and Design*, *80*(January), 31–44.

2. Kacker, R., Regensburg, S. I., & Kramer, H. J. M. (2017). Residence time distribution of dispersed liquid and solid phase in a continuous oscillatory flow baffled crystallizer. *Chemical Engineering Journal*, *317*, 413–423.

3. Dickens, A. W., Mackley, M. R., & Williams, H. R. (1989). Experimental residence time distribution measurements for unsteady flow in baffled tubes. *Chemical Engineering Science*, *44*(7), 1471–1479.

4. Ni, X., & Pereira, N. E. (2000). Parameters Affecting Fluid Dispersion in a Continuous Oscillatory Baffled Tube. *AIChE Journal*, *46*(1), 37–45.

540773 2018 Outlook for Energy: A View to 2040

Abstract:

The Outlook for Energy is ExxonMobil's global view of energy demand and supply through 2040. It also highlights the dual challenge of providing affordable energy to support prosperity while reducing environmental impacts – including the risks of climate change. We use the Outlook to help inform our long-term business strategies and investment plans.
The Outlook anticipates significant changes through 2040 across the world to boost living standards, reshape the use of energy, broaden access to abundant energy supplies, and accelerate decarbonization of the world's energy system to address the risks of climate change.

- Global energy needs will rise about 25 percent over the period to 2040
- Electricity demand will nearly double in non-OECD countries
- Worldwide electricity from solar and wind will increase about 400 percent
- Natural gas will expand its role, led by growth for electricity generation
- Growth in oil demand will be driven by commercial transportation and chemical needs
- Efficiency gains and growing use of less-carbon intensive energy sources will contribute to a nearly 45 percent decline in the carbon intensity of global GDP

Energy is fundamental to modern life, and as the world's population approaches 9 billion people in 2040, we are challenged to improve living standards everywhere. Understanding the factors that drive the world's energy needs – and likely choices to meet those needs – is the mission of the Outlook. By sharing the Outlook with the public, we hope to broaden that understanding. Energy matters to everyone, and we all play a role in shaping its future.

541543 Energy Decarbonisation Scenarios

Abstract:

With the energy sector being associated with 2/3 of the annual greenhouse gas emissions, decarbonizing the energy is a key step to reduce their level. The National Determined Contributions (NDC's) announced so far by the participating countries do not meet the objectives of the Paris-climate agreement, as far as reaching a peak in emissions in the near future or decreasing to a level that is compatible with the stated long-term targets.

Using different scenarios for energy demand and supply, options for energy sector decarbonisation are evaluated. Meeting the 2-degree target requires a major transformation in the power sector, efficiency improvements – specially in the buildings and in the industry, and a revolution in transport towards fuel and modal shifts. With a significant increase in renewable energy, system integration and optimization will also be key in reducing the cost of the energy transition. Meeting more ambitious climate goals ("well below 2 degrees") will push further the technological innovation and implementation requirements throughout the entire energy value chain. A scenario aiming at zero net-emissions by 2060 is presented, along with a comparison to the 2-degree case.

542625 Fundamental Research Needs to Advance Energy Technologies

Abstract:

The U. S. Department of Energy (DOE), Office of Science, Office of Basic Energy Sciences (BES) supports fundamental research in chemical and materials sciences to provide the foundations for new energy technologies and to support DOE missions in energy, environment, and national security. This presentation will discuss opportunities for fundamental research to impact DOE's energy mission "to catalyze the timely, material, and efficient transformation of the nation's energy system and secure U.S. leadership in energy technologies" with a focus on the way we generate, store and use energy nationally. I will provide an overview of BES strategic planning over the past decade that identified priority research directions for advancing energy applications, highlight key scientific advances in these areas, and discuss some future opportunities for modern science, particularly chemical sciences, to accelerate the transformation of the U. S. energy portfolio.

529993 Experimental and CFD Studies of a New Continuous Process for Mixing of Complex Non-Newtonian Fluids

Abstract:

Experimental and CFD studies of a New Continuous Process for Mixing of Complex Non-Newtonian Fluids

Simona Migliozzi[1], Robert Sochon[2], Luca Mazzei[1] and Panagiota Angeli[1]

[1]*Department of Chemical Engineering, University College London, Torrington Pl, London WC1E 6BTA, UK*
[2]*GlaxoSmithKline Consumer Healthcare, St George's Avenue, Weybridge, KT13 0DE, UK*

1. Introduction

Mixing processes involving non-Newtonian fluids are widely employed in several industrial applications. Specifically, the design of new continuous mixing operations poses many challenges, especially when dealing with highly non-Newtonian fluids. Typically, batch processes are employed to perform blending in high viscosity conditions. However, the large fluid volumes that characterize these operations can lead to concentration and temperature gradients within the vessels and to the establishment of dead mixing zones for more viscous fluids. New continuous processes can prevent these problems and achieve better performance at lower costs. In particular, static mixers are a promising alternative, especially for blending highly-viscous fluids that can be processed exclusively in laminar flow conditions. Indeed, in this flow regime, static mixers can enable radial mixing via a periodic sequence of splitting and recombining of the fluid streams that reduces progressively the thickness of each fluid layer, eventually promoting diffusive mass transfer between the two phases [1].

In this work, our main objective is investigating the mixing efficiency of two different static mixers and the effects of the rheological properties of the mixture using experimental and computational fluid dynamics (CFD) methods. Our focus application is the manufacturing of a newly formulated non-aqueous toothpaste. Therefore, a complete characterization of the complex rheological behaviour of the specific working mixture was also carried out at different temperatures and compositions.

2. Materials and methods

The two liquid phases employed in the process of interest are glycerol and a carbomer suspension in polyethylene glycol. When mixed, the liquids start forming a gel. We studied the structure evolution of the liquid mixture using time-resolved rheometry [2-3] and estimated the gelling time at different temperatures and glycerol mass fractions. Then, the flow curves of the final gels were obtained for the same ranges of temperature and mass fractions. A TA Instruments Discovery Hybrid Rheometer, equipped with a Peltier plate to precisely control the operating temperature, was used to carry out all the rheological measurements.

To perform the mixing studies, two different static mixers were selected: the Kenics, which has been extensively studied in the literature, thus allowing a first comparison with literature correlations [4-5], and the GFX, which is a modification of the SMX (Sulzer technology). In both geometries all the elements were detachable, thus allowing to easily change the overall length of the mixer. Pressure drop measurements and Planar Laser Induced Fluorescence (PLIF) techniques [6] were used for the experimental campaign. A schematic of the whole experimental apparatus is shown in Figure 1. ANSYS Fluent 18.2 was used to perform the simulations. Specifically, we selected the 'Species transport' model, present in the software, given the miscibility of the two fluids.

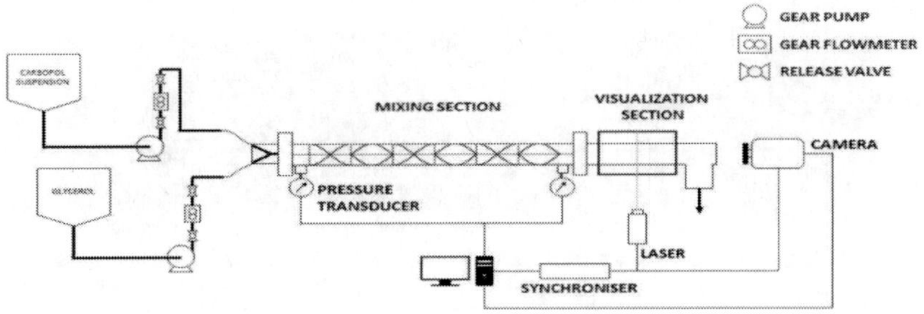

Figure 1. *Schematic of the experimental apparatus.*

3. Results

The flow properties of the activated mixtures show a clear dependence of the mixture viscosity on both composition and temperature (Figure 2-3). Therefore, a viscosity model that accounts for this dependence should be added in the Species transport model of Fluent.

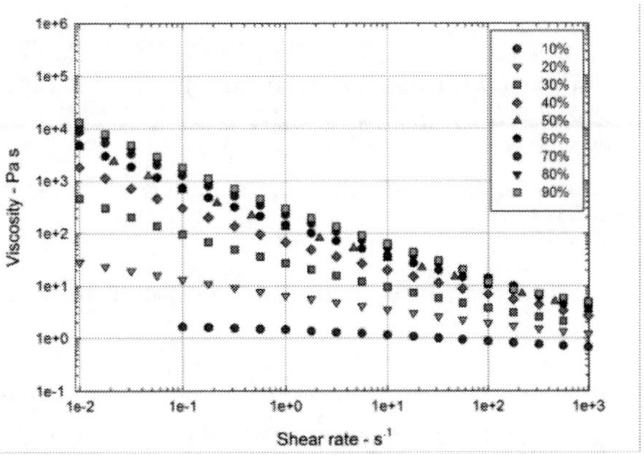

Figure 2. Flow curves of the activated mixtures for all mass fractions of Carbopol suspension studied at a fixed temperature.

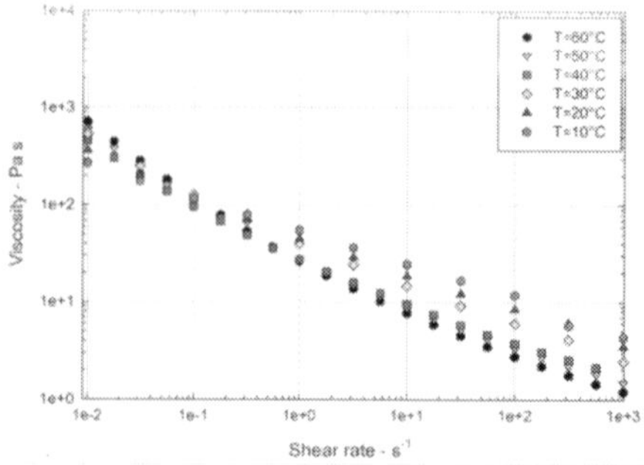

Figure 3. Flow curves of the activated mixtures for all mass fractions of Carbopol suspension studied at a fixed mass fraction of glycerol in polymer suspension.

Pressure drop values and concentration maps were obtained from the experimental apparatus and compared with the CFD results. Thanks to the possibility to remove the elements, we were able to compare the concentration fields at different cross-sections of the computational geometry (Figure 4) with the results obtained from the PLIF experiments. The main differences found between the experimental and computational concentration profiles can be addressed tothe presence of numerical diffusion in the CFD results, which is a non-physical diffusion phenomenon that arises from the discretization of the nonlinear convective term in the Navier-Stokes equations. Therefore an analysis of numerical diffusion was carried out to understand the phenomenon [7].

Figure 4. *Volume fraction evolution in the GFX static mixer geometry.*

References

[1] *Muzzio, F.J., Swanson, P.D., Ottino, J.M.,* The statistics of stretching and stirring in chaotic flows. *Phys Fluids A 3, 822, 1991.*

[2] *Holly, E. E., Venkataraman, S. K, Chambon, F., Winter, H. H.,* Fourier transform mechanical spectroscopy of viscoelastic materials with transient structure. *J. Non-Newton. Fluid., 27, 17–26, 1988.*

[3] *Curtis, D. J., Holder, A., Claypole, J., Walters, M., Thomas, B., Barrow, M., Deganello, D., Brown, M. R., Williams, P. R., Hawkins, K.,* Validation of Optimal Fourier Rheometry for rapidly gelling materials and its application in the study of collagen gelation. *J. Non-Newton. Fluid., 222, 253–259, 2015.*

[4] *Meijer, H.E.H., Singh, M.K., Anderson, P.D.,* On the performance of static mixers: A quantitative comparison. *Prog. Polym. Sci. 37, 1333 – 1349, 2012.*

[5] *Shah, N.F., Kale, D.D.,* Pressure drop for laminar flow of non-Newtonian fluids in static mixers. *Chem. Eng. Sci. 46, 2159-2161, 1991.*

[6] *Alberini, F., Simmons, M.J.H, Ingram, A.,* Use of an areal distribution of mixing intensity to describe blending of non-Newtonian fluids in a Kenics KM static mixer using PLIF. *AIChE J. 60, 332 – 342, 2014.*

[7] *Liu, M.,* Computational study of convective-diffusive mixing in a microchannel mixer. *Chem. Eng. Sci. 66, 2211 – 2233, 2011.*

530218 Elastic Flow Compartments in Stirred Tanks

Abstract:

Elastic flow compartments in stirred tanks

Markus Kolano, Matthias Kraume, TU Berlin, Chair of Chemical and Process Engineering, Ackerstraße 76, 13355 Berlin, Germany

Mixing processes in stirred tanks in which one or more liquids exhibit viscoelastic flow properties are of particular importance in food, polymer and biotechnology. It is well known that elasticity can lead to flow phenomena like the Weissenberg effect or flow field inversion when stirring with a Rushton turbine: While at low rotational speeds a completely axial secondary flow field can be formed and elastic forces dominate, an increase in stirring rate and thus in centrifugal forces finally results in a radial flow profile. Between these extreme cases, cavern like flow compartments, that is to say almost completely separate circulation vortices with limited convective exchange, form, which result from mutually competing elastic and centrifugal forces. These elastic compartments can form using any stirrer geometry and have a significant effect on the mixing efficiency of the stirred tank (see figure 1). Knowledge of their existence is therefore of high interest.

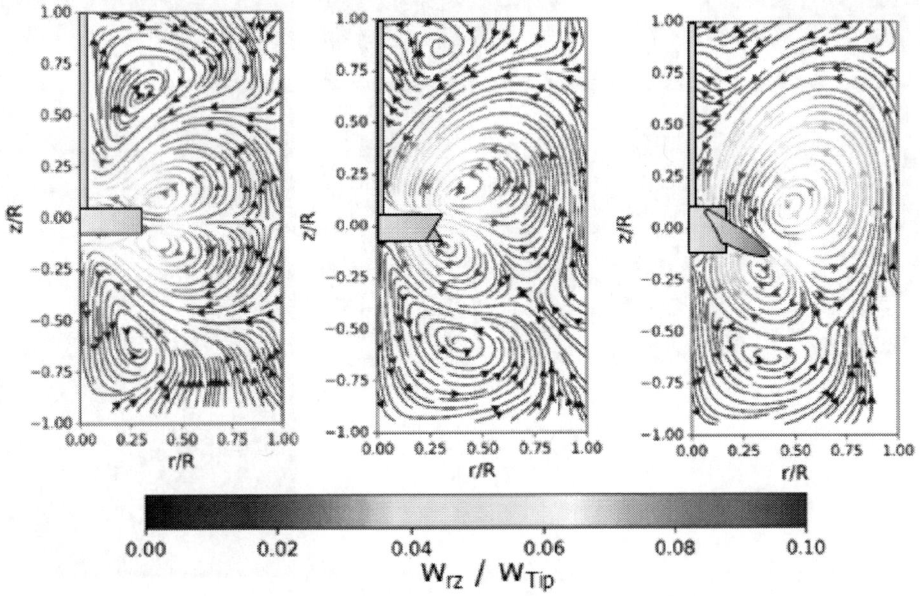

Figure 1: PIV measurements of secondary flow fields showing elastic flow compartments; left: Rushton turbine in aqueous xanthan gum solution (1 g/kg), middle and right: pitched blade turbine and 3-blade propeller in aqueous CMC solution (10 g/kg)

In this presentation, the influence of elastic flow properties as well as the rotational speed on the formation of these flow compartments is investigated. Secondary flow fields are presented for different stirrer geometries (Rushton turbine, pitched blade turbine and propeller stirrer), which were obtained by the PIV method for various viscoelastic fluids (aqueous xanthan gum and carboxymethyl cellulose solutions). In addition, the pLIF method is used to investigate the mixing behavior of these systems and to analyse the extent to which convective mass transport takes place between the compartments as well as their sizes and forms (see figure 2). It can be shown that by using the elasticity number a general classification of flow regimes in a transition state between elastic or inertia dominated flow for all analysed geometries and fluids is possible. These results will be supported by CFD simulations using viscoelastic constitutive equations.

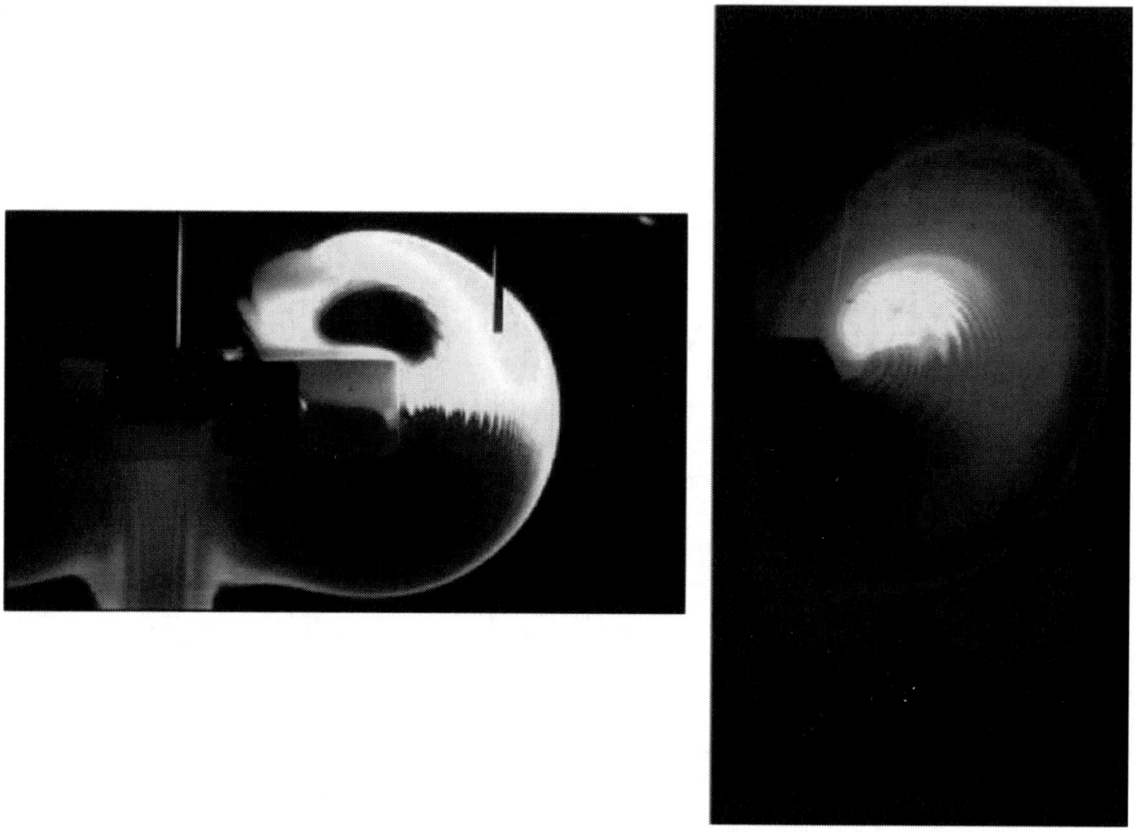

Figure 2: pLIF measurements; left: detailed study of a flow compartment caused by Rushton turbine, right: flow compartment caused by a pitched blade turbine

530352 Flow of Fluids with Evolving Rheology in Open Pipes and Static Mixers

Abstract:

In many industrial applications fluids undergo rapid changes in rheology due for example to polymerization, heat transfer or other transformations. To adequately design continuous processes involving such flows one must account for the coupled effects between rheology kinetics and the generally non-uniform residence times. In this work we investigate the relation between pressure drop, residence time distribution and rheology kinetics and describe the application of various approaches including computational fluid dynamics, experiments, reduced order models and correlations using dimensionless numbers.

540965 Optimizing Agitation in Suspension Polymerizers

Abstract:

Suspension polymerization is a common process for making products such as PVC. Droplets of monomer are dispersed in water in the presence of a surfactant and once the desired droplet size is achieved an initiator is added to start the reaction. The size of the polymer particles produced is governed by the size of the monomer droplets formed at the start of the process.

As the reaction proceeds, coagulum is formed which can coat the impellers and vessel walls or form large masses which can damage the agitator if they impact the blades or shaft. The formation of coagulum can be minimized by using an impeller that generates a high flow but, at the same time, it must provide sufficient shear to generate the desired monomer droplets.

This presentation will compare impeller properties and review how they can be selected to achieve the correct balance to flow and shear. A case study on retrofitting in a suspension polymerization reactor will be presented.

533914 Influence of Mixing Rate and Temperature during the up-Scaling of Emulsified Cosmetic Products

Abstract:

Global cosmetics market has been steadily growing during last years, and the most recent forecasts indicate that this tendency will remain. A major driver of this market boost is the high rate at which the new and innovative products are requested by the more sophisticated consumers. Then, in some cases, the minimization of the time required to introduce a product into the market is more important than an optimized product performance. In spite of the current availability and continuous development of tools for product conception and design, together with the availability of high throughput prototyping and manufacturing techniques, a major limiting factor remains in the up-scaling process. Specifically, in the development of structured products in the cosmetics industry (e. g. emulsions), up-scaling to industrial production is a time-consuming, resources-intensive, and complex process; reason why it frequently must incorporate trial and error techniques.

During the manufacture of cosmetic emulsions, several variables need to be considered to ensure obtaining the desired properties of the product and long term stability. Variables such as emulsification temperature, mixing times, stirring rates, equipment geometry, ingredients addition procedures, heating or cooling rates; all these play a major impact in the final characteristics of the emulsified products. During the up-scaling process, these variables need to be tuned and adjusted along the way, mainly through trial and error methods. This makes the process highly inefficient.

Alternatively, different systematic up-scaling techniques have been developed for cosmetics production. Specifically, for water-in-oil emulsions, the use of process invariants such as some dimensionless numbers of the form $ND^{Î±}$, have been incorporated as key parameters during the up-scale process. While N and D relate geometries and flow regimes used in the emulsification process, Î± is an adjusted parameter that largely depends on the nature (properties, compositions, specific synergies) of the mixture. Even when up-scaling with different equipment geometries, stirring rates, of operating policies, the same final product is expected to be obtained by maintaining constant such process invariants among scales.

Nevertheless, most emulsified cosmetics are oil-in-water emulsions, and the up scaling process of this type of products via process invariants has been seldom studied. In this regard, this work studied the influence of some process variables (emulsification temperature, mixer type and speed, and cooling rates) in the production of an oil-in-water cosmetic emulsion. The impact on the major characteristics of the product (drop size distribution and rheological behavior) were assessed to define the corresponding process invariants during up-scaling. Experiments were carried out at different scales (from 100mL to 1 L) using rotor-stator mixer operating under high stirring rates (3000 to 10000rpm), an anchor mixer in rates among 50rpm and 200rpm, and emulsification temperatures between 70°C and 90°C. The ingredients of the cosmetic product corresponded to the base emulsion of a body cream. As a result of the study, process invariants for oil-in-water emulsions preparation were identified, and they were successfully validated in the up-scaling of the process to a larger batch (10 L).

Abstract:

Experimental and computational studies of the fluid dynamic behaviour of liquid-solid mixtures in agitated vessels

Giovanni Meridiano, Weheliye Hashi Weheliye , Luca Mazzei, Panagiota Angeli

Department of Chemical Engineering, University College London, Torrington Pl, London WC1E 6BTA, UK

giovanni.merigiano.17@ucl.ac.uk; weheliye.weheliye.10@ucl.ac.uk; l.mazzei@ucl.ac.uk p.angeli@ucl.ac.uk, ,

Introduction

Mixing of solids in viscous liquids in stirred vessels is a crucial step in many manufacturing processes; it is regularly encountered in a wide range of industrial sectors like health-care, pharmaceuticals, cement manufacturing and food processing. In many of these applications the liquids have complex non–Newtonian behaviour (Paul, Atieno-Obeng and Kresta, 2004). In these industrial processes, obtaining a uniform mixture is a difficult task, because there are no turbulent eddies to help distribute the components.

The objective of this work is to understand and characterize the fluid dynamics of uniform liquid-solid mixtures with both Newtonian and non–Newtonian liquid matrixes in stirred vessels. For this purpose, we measured the power consumption and the velocity profiles at different solid volume fractions and compared the results with those obtained via Computational Fluid Dynamics (CFD) simulations.

Materials and methods

The power required for stirring the suspensions as well as the internal velocity fields obtained via Particle Image Velocimetry (PIV) were measured experimentally at different solid volume fractions in a stirred vessel.

The measurements were carried out in a stirred tank specifically designed to reproduce real industrial scale equipment. It consists of a transparent cylindrical tank of internal diameter D_T= 50 mm equipped with two baffles and a dual-blade impeller of diameter D = 37 mm. The height of the fluid, H, in the tank was set equal to the tank diameter. The impeller was located at the centre of the tank, 10 mm from the bottom, and was driven by a variable speed motor that could operate in the range of 50–2000 rpm (IKA Eurostar 20) (Fig. 1). As viscous Newtonian fluid, we employed a solution of corn syrup and water, while, as non–Newtonian fluid, we opted for a solution of corn syrup, water and xanthan gum. The solid in both cases is PMMA microspheres (32–37 μm).

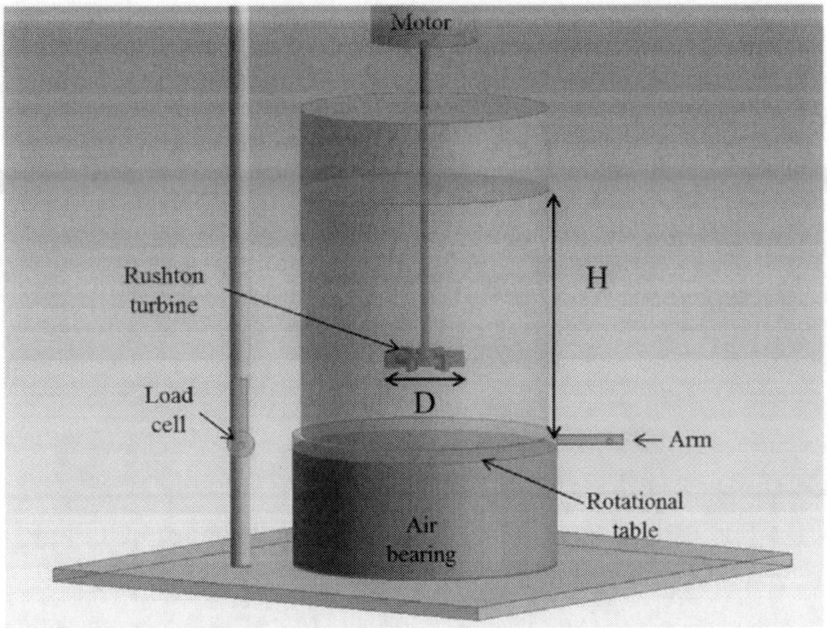

Figure 1: Experimental setup for the measurement of power consumption in the vessel

Power consumption measurements

A precise way to measure power consumption is to use an air bearing system. A schematic of the air bearing system used in this work is shown in Fig. 1. From the force that is required to stop the rotation of the rotational table the power required to agitate the fluid can be calculated as follows:

$$P = 2\pi N M_I$$

where M_I denotes the axial torque applied to the fluid by the impeller and N is the impeller angular speed in rotations per unit time. We recorded this force with a load cell (Omega LCM601-1) and a data acquisition system and software (OmegaIN-USBH).

PIV measurements

Velocity field measurements were carried out using Particle Image Velocimetry. For these experiments, we used the same transparent tank used for the torque measurements, but in this case, the tank was enclosed in a square transparent box. This box was made of acrylic and filled with a refractive index matching fluid to avoid optical distortions on the surface of the cylindrical vessel. The PIV set-up includes a dual cavity Nd:Yag green laser (532 nm) (Litron Laser®, 15 Hz, 1200 mJ) and a straddling CCD camera with 1280 x 1024 pixels (TSI PowerView™ Plus) that operates at a maximum frequency of 15 frames per second. The camera is equipped with an AF Nikkor 50mm f/1.8D prime lens (Nikon®). A hall switch sensor was used to capture images at the same phase angle. As tracers we used fluorescent polymer particles (melamine resin based) coated with rhodamine B (20µm), which absorb at 532 nm and emit at 610 nm. They are neutrally buoyant in the fluids considered and, at the experimental conditions explored; their relaxation time is negligible compared to the convection time. The laser and the camera were synchronized by means of a Laser Pulse Synchroniser (Model 610035 TSI) and were controlled via the Insight 4G (TSI) software. The laser beam passed through a collimator (Model 610026 TSI) and two cylindrical lenses (25 mm, and 15 mm) to create a narrow plane of 1 mm thickness. The laser plane was then reflected on a 45° silver coated mirror and entered the stirred vessel from the bottom. A sketch of the setup is shown in Figure 2.

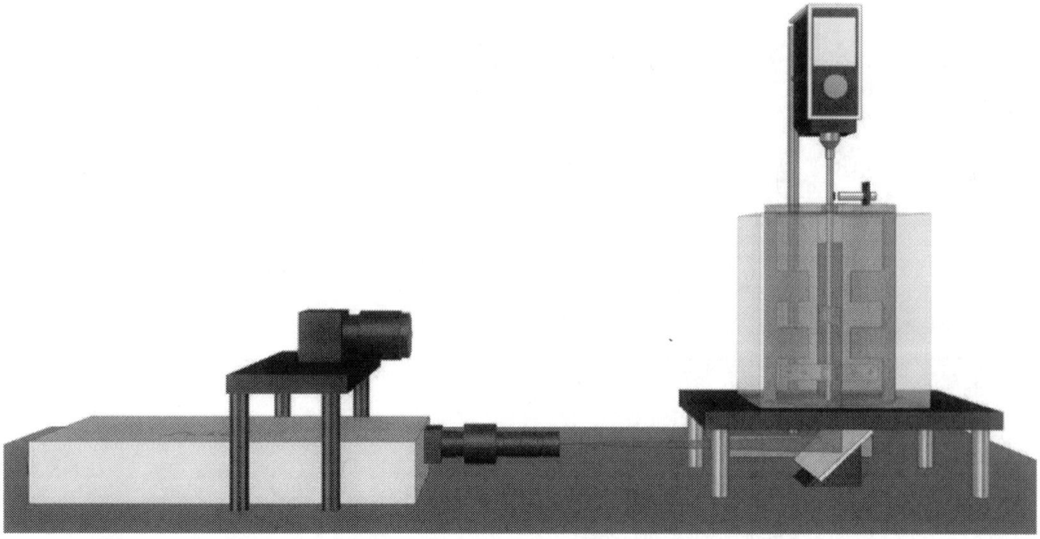

Figure 2: Sketch of the main components of the experimental set-up for the PIV measurements

Computational fluid dynamics simulations

Computational Fluid Dynamics is a powerful tool that yields relevant process information that can be used to design and assess the performance of stirred vessels. A number of researchers have used CFD to study the behaviour of different types of liquid-solid systems in mechanically stirred tanks with different types of impellers (Blais *et al.*, 2016) (Konz and Windhab, 2016) (Fradette *et al.*, 2007). All the previous studies highlighted the importance of validating the CFD models against experimental data.

The model employed in this study follows a *Eulerian–Eulerian* approach. This means that the transport equations are written in term of volume–averaged quantities. For the mixture stress closure an experimental constitutive equation is implemented that correlates the suspension viscosity to the solid volume fraction. Therefore, the first step of our work focussed on the rheological characterization of the two suspensions mentioned in Section 2 and on the development of suitable rheological constitutive equations for the CFD model.

We compared the CFD findings against the power measurements in the stirred vessels and the velocity fields for different impeller speeds and solid loadings.

Preliminary Results

We characterized the rheology of the Newtonian fluid–particles system, at different solid volume fractions, using a HR-3 Discovery Hybrid Rheometer (TA Instruments®). Some results are shown in Fig.3.

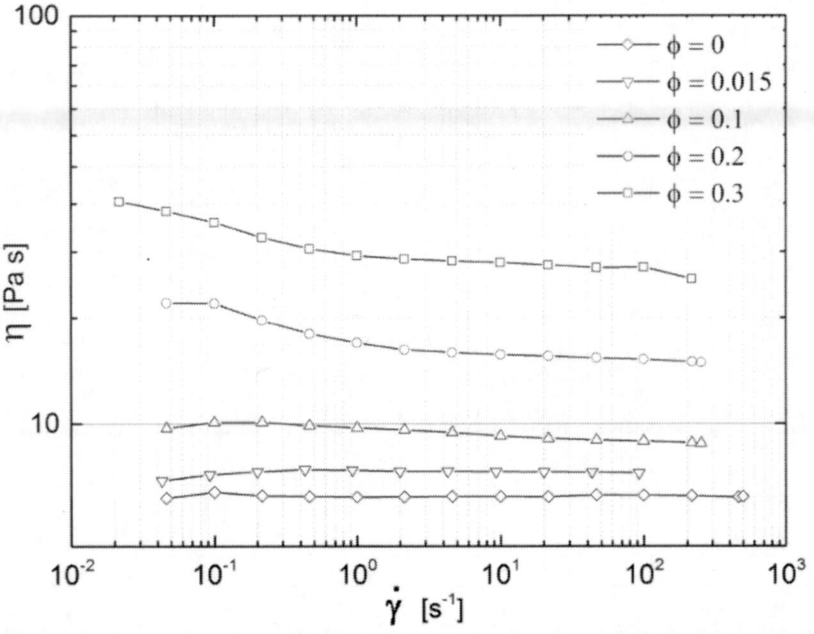

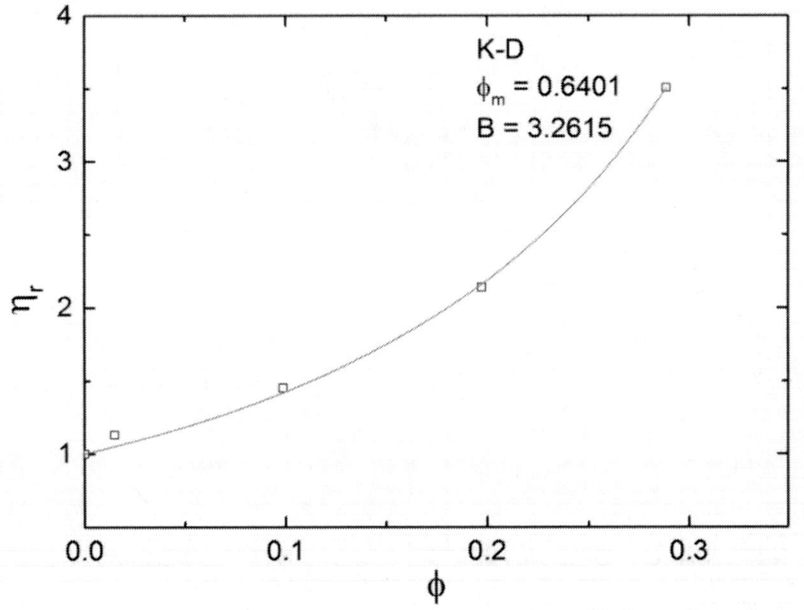

Figure 3: Evolution of viscosity with solid volume fraction F

As expected, the viscosity increases with the solid volume fraction; moreover, the overall behaviour of the suspension remains Newtonian at solid volume fractions below 0.2 while at higher solid volume fraction the suspension shows a slightly shear-thinning behaviour. This phenomenon has been widely documented in the literature (Stickel and Powell, 2005). In addition, the change in viscosity with volume fraction is well predicted by the Krieger–Dougherty equation, whose parameters assume values similar to those found in the literature.

The power consumption for the Newtonian fluid alone has been measured to calibrate the instruments and verify the reproducibility of the results. Furthermore, we ran some simulation for the same system. In Fig. 4, we report both the experimental and computational power numbers against the impeller Reynolds number.

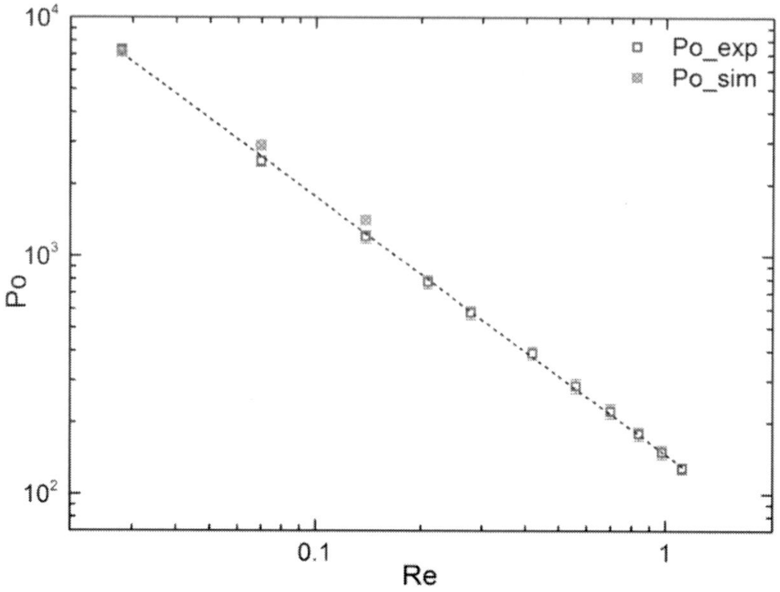

Figure 4: Power number against Reynolds number for the Newtonian fluid–particles system

We were able to recover the well-established relation between the power number and the Reynolds number in the laminar regime (Paul, Atieno-Obeng and Kresta, 2004). Moreover, there is good agreement between the experimental and computational results. <>

Bibliography

Blais, B. *et al.* (2016) 'Development of an unresolved CFD-DEM model for the flow of viscous suspensions and its application to solid-liquid mixing', *Journal of Computational Physics*. Elsevier Inc., 318, pp. 201–221. doi: 10.1016/j.jcp.2016.05.008.

Fradette, L. *et al.* (2007) 'CFD phenomenological model of solid-liquid mixing in stirred vessels', *Computers and Chemical Engineering*, 31(4), pp. 334–345. doi: 10.1016/j.compchemeng.2006.07.013.

Konz, A. K. and Windhab, E. (2016) 'Experimental and computational study of a high speed pin mixer via PEPT, visualization and CFD', *Chemical Engineering Science*. Elsevier, 155, pp. 221–232. doi: 10.1016/j.ces.2016.08.007.

Paul, E. L., Atieno-Obeng, V. A. and Kresta, S. M. (2004) *Handbook of Industrial Mixing*, *Wiley Online Library*. doi: 10.1002/0471451452.

Stickel, J. J. and Powell, R. L. (2005) 'Fluid Mechanics and Rheology of Dense Suspensions', *Annual Review of Fluid Mechanics*, 37(1), pp. 129–149. doi: 10.1146/annurev.fluid.36.050802.122132.

522152 Optimizing Productivity of Bioreactors with Multiphase Simulation on GPUs

Abstract:

Computer simulation has become a vital tool in learning to understand the processes inside multiphase reactors such as aerated stirred tanks used in biopharmaceutical manufacturing. This in-silico design approach allows for the identification of bottlenecks that limit the overall efficiency of large scale bioreactors, without relying on empirical knowledge gained on a pilot-plant.

In order to simulate the three dimensional flow field inside multiphase reactors in an acceptable wall time, a highly-parallelized implementation of the lattice Boltzmann method is performed on state-of-the-art graphics processing units. A Lagrangian model is used to describe the movement of air bubbles inside the reactor, including coalescence and break-up effects. The phases are two-way coupled. Advection of substrate, oxygen and carbon dioxide is modelled by a novel species transport model that takes advantage of the high degree of parallelization.

An additional carbon source in the form of emulsified oil drops is modelled by locally solving a population balance equation including a coalescence and break-up kernel. The volume fraction of oil droplets affects the local fluid mixture viscosity. Local parameters such as oil, oxygen and substrate concentration as well as shear rate are used in a biological model that predicts the productivity of individual biological cells. These cells are introduced as massless particles which follow the fluid flow field. The trajectory and age of each cell inside the reactor affect the biological model.

Simulations are made for several bioreactor sizes ranging from lab scale up to the full industrial production scale. The results are compared to literature data. Additionally, the gas phase phase movement, holdup and gas absorption is validated with measurements made on a 150l pilot-plant reactor.

535754 Flow inside Bioreactors: Comparing Predictions from Lattice-Boltzmann to Experimental Data

Abstract:

The lattice-Boltzmann method (LBM) provides an inherently time-dependent approach for modeling momentum transport through open and closed bioreactors. At low Reynolds numbers involving laminar and transitional flow, LBM can be used to perform direct numerical simulations that resolve the full spectrum of eddies inside the reactor. In turbulent systems, LBM can be paired with a sub-grid eddy viscosity model to perform large eddy simulations (LES) that resolve those eddies larger than the local grid spacing. These insights into flow turbulence have important implications on the design of bioreactors, as related to cell growth, product homogeneity, and processing time.

In this work, we examine agitated tanks and compare the mean and time-varying flow fields predicted from LBM to those obtained from PIV and LDA measurements. In regions both within and away from the trailing vortex, we find good quantitative agreement between the predictions from LBM and the experimental data. Next, we assess the effects of system resolution on the predicted flow field and the coupling between lattice spacing, energy dissipation, and the effects of the sub-grid filter. We then show how this technique can be used to optimize bioreactor performance, as characterized by blend time and energy dissipation.

536050 CFD Modeling of Oxygen Dissolution in Bioreactors: Mass Transfer and Population Balance Study in Stirred Tanks

Abstract:

Aerobic microorganisms need oxygen for sustenance, growth and product formation. Furthermore, a continuous supply of oxygen is critical in aerobic bioreactors due to its low solubility in aqueous solutions. While maintaining continuous oxygen transfer from air to the broth is necessary for the optimum design and scaling of these reactors, it is a rather complex task. The intricacies commonly encountered in the design of these systems are highly influenced by various parameters including the geometry, operating conditions, turbulence, gas distribution among others. A typical configuration for bio-reactors encountered in the bio-pharmaceutical and fine-chemical manufacturing industry, is gas-liquid dispersion in stirred tanks. Most studies encountered in the literature focus on the separate analysis of gas holdup, bubble size distribution, and mass transfer coefficients. However only a few thus far have considered using a complete set of robust CFD models at operating conditions (Nallamothu et al., 2015).

The current work extends the research presented by Nallamothu (Nallamothu et al., 2015), by including more detailed physics such as a swarm factor drag correction for the Ishii-Zuber drag coefficient (Ishii et al., 1979) which takes into account the presence of clusters of bubbles. A parametric study has also been conducted to account for the effect of the particle size distribution using the Inhomogeneous Discrete population balance model in the CFD software ANSYS/FLUENT. In contrast to assuming the same velocity field for all particle sizes within a phase, this approach models the particle size distribution across multiple phases which can realistically predict the segregation of large and small bubbles.

The current study considers different rotational speeds and gas flow rates. The mass transfer coefficient is obtained by experimental measurement of dissolved oxygen concentration by absorption in pure water. Finally, the simulation results are validated against experimental data published by Laakkonen et al. (Laakkonen et. al., 2007) in terms of temporal variation of normalized concentration as well as variation of gas holdup with gas flow rate.

References:

1) Nallamothu, S.K.; Ozarkar, S.; Joshi, S. "Advances in Gas Sparging Simulation for Bioreactor Modeling to Create Comprehensive Simulation Design Space". AIChE Annual Meeting (2015).

2) Ishii, M.; Zuber N. "Drag Coefficient and Relative Velocity in Bubbly, Droplet or Particle Flows". AIChE J. 25: 843-855 (1979).

3) Laakkonen, M.; Moilanen, P.; Alopaeus, V.; Aittamaa, J. "Modelling local bubble size distribution in agitated vessels". Chem. Eng. Sci. 62: 721-740 (2007).

537906 Kinetic Study and CFD Modeling of Anaerobic Bioethanol Fermentation

Abstract:

Nowadays, computational modeling of biochemical and biological systems plays an important role in biotechnology-related disciplines. In this context, mathematical models and process modeling techniques are frequently employed to predict and describe the behavior of bioprocesses such as fermentation. One of the challenges in bioreactor modeling is interaction of mass transfer and biochemical reaction in bio processes. A MATLAB-based graphical user interface (GUI) was developed to study the kinetics of bioethanol production and yeast growth on glucose fermentation using seven different Monod type models. This model-based application can be of assistance in estimating and developing the procedures involved in upscaling and optimizing the bioprocesses of interest. In this study, an anaerobic batch ethanol fermentation carried out in a 3-L bioflo115 with two Rushton turbines was considered. The developed kinetic application has produced acceptable results compared to experimental data obtained in laboratory scale ethanol fermentation. The estimated parameters were applied in CFD study of the process using the k-Îµ turbulent model to consider the effect of mixing on fermentation productivity. Fermentation productivity and concentration of substrate, biomass and product predicted in both perfectly-stirred condition and CFD model were compared to each other and to experimental data obtained for ethanol produced, dry yeast Saccharomyces cerevisiae growth and glucose degradation in laboratory scale fermentation.

540731 Bubble-Scale Modeling of Gasified Reactors

Abstract:

Aerobic and microaerobic fermentation processes require significant aeration for proper organism performance in both stirred tanks and air-lift columns. Aeration provides oxygen for fermentation as well as mixing for the bulk fermenter. Air flows are high and have a significant impact on the overall fluid flow. In the case of air-lift columns, air flow is the only mechanism for fluid flow.

Fermentation productivity is often limited by the rate of oxygen transport into the system, characterized by the bulk mass transfer coefficient $k_L a$, a product of the local mass transfer coefficient k_L and the interfacial area a, both strongly dependent on the turbulent dissipation. Both terms are difficult to predict by standard CFD methodology.

In this work, we use lattice-Boltzmann simulations to model gasified reactors with a two-way coupling between the fluid and gas bubbles. We begin by showing how it is practical to track the trajectories of hundreds of millions of individual bubblesâ€"populations that are comparable to the total number of bubbles within typical industrial-scale gasified systems. Next, we review experimental correlations used to predict the coalescence behavior of bubble pairs, as a function of the approach velocity and associated Weber number. We then review experimental data related to bubble break-up, as a function of fluid shear and power dissipation. Then, by superimposing these relationships on to the explicit bubble trajectories/collisions captured in the simulation, we model and simulate the real-time, spatiotemporal variation in bubble diameter. Finally, we use various approaches to connect bubble size and turbulence to $k_L a$. We compare these predictions to available experimental data, including gas hold-up, gasified power draw, and mass transfer rates.

542620 The Impact of Shale Gas and Oil on the Chemical Industry

Abstract:

Lower alkanes and aromatics are very useful as both fuels and chemical raw materials. Occasionally supply disruptions constrained availability but also encouraged investigation of alternative feedstocks and chemistries. Recent new technologies for the production of unconventional gas and oil sources from tight shale formations significantly increased availability of both methane and condensate, radically altered process economics, resulted in the construction of much new intermediates capacity, and multiplied the ways traditional and alternative chemistries might be exploited. This presentation will outline historically how natural gas components became useful as chemical feedstocks, how government regulations occasionally led to perturbations in availability and price, how supply crises led to the development of alternative chemistries, catalysts, and ultimately the exploitation of shale resources. More specifically, will be discussed how excess coproducts, especially condensate, from increasing shale gas and oil production for energy are having significant unexpected impacts on the chemical process industries and opportunities for new process chemistry research.

541186 Sustainable Energy and Chemicals: Past, Present, and Future

Abstract:

The last 40 years has seen the U.S. energy situation vacillate from gasoline rationing and concerns over Peak Oil, to projected energy and feedstock prices that were noncompetitive in global markets for chemicals production, to the economic revolution of frack gas and light tight oil. The importance of low cost energy and feedstocks in driving economic opportunity will be discussed relative to this history, along with the importance of sustainability in decision making to gain stakeholder buy-in and acceptance. Strategies for managing career paths and competencies despite changing targets and high uncertainty will be addressed.

542618 Disruptions: What the Future May Hold

Abstract:

Disruptive forces are all around us. They change the way we communicate and work. They change the way we develop goods and services. They also change the types of products that are manufactured. Examples such as the way in which we produce and consume energy or the feedstocks we use to produce product have seen many changes over the course of the last few decades, and these disruptions are expected to continue. The chemical engineering discipline and the expectations for a chemical engineerâ€™s skills has also changed and adapted as a result of these disruptive forces. A discussion of the broad themes associated with these forces and the implications that they may have to chemical engineers in the future will be discussed.

542621 Geopolitical Factors Influencing the Evolution of the Chemical Industry

Abstract:

The basic chemicals and plastics industry is built on a simple principle: find a cheap feedstock and convert it into something more valuable. However, the right feedstock depends on place and time. Certain policies and actions of various governments have influenced the evolution and profitability of the chemical industry during the past 50 years, sometimes in surprising ways. For example, high oil prices and US government funding encouraged the development and commercialization of coal gasification technology during the late 1970s and early 1980s. This enabled China to develop a competitive basic chemical industry using coal as a feedstock during the last 25 years. Similarly, an abrupt increase in oil prices during the first decade of the 21st century encouraged the present shale gas revolution, causing a major shift in the economy of producing basic chemicals and polymers in the US. If history is our guide, then what governments choose to fund and regulate may be as important as market forces in determining the shape of our industry and profession in the future.

542622 Agility & Resilience: How to Maintain Career Competitiveness in the Changing Chemical Industry

Abstract:

The world around is changing. The shift to protectionism, global instability and price volatility of raw materials and feedstocks has the potential to wreak havoc on the chemical industry. With the revolution of shale gas in North America and the sudden drop in feedstock costs, the chemical sector in the US has reasserted its historic dominance. Learn how the role of a chemical engineer can contribute to industry growth and resilience, while adapting to allow for personal agility throughout the career life cycle.

529812 CFD Modeling of Turbulent Reactive Flows with Bourne Chemistry Comparisons

Abstract:

Reactive Computational Fluid Dynamics (CFD) simulation can be a powerful tool in designing industrial mixers. Many industrially-relevant reactive flows involve multiple parallel, series and competing reactions of different time scales. Slow reactions (Damköhler number<<1) are limited by the kinetics rate. Faster reactions (Damköhler number >>1) are limited by the mixing rate of the reactants. An accurate representation of the key competitive reactions as well as the turbulence effect, in a CFD model or during experiments using a model chemistry, is of key importance for mixer/reactor design.

This presentation aims to both highlight how these issues have been addressed from an industrial standpoint as well as expose the technological gaps that exist in the current experimental and modeling tools. The study case consists of a converging jet in cross-flow and includes a kinetic scheme with series-parallel reactions. A laboratory representation of a portion of the industrial mixer/reactor was constructed and reactive (the fourth Bourne chemistry: 2, 2 DMP and caustic neutralization with acid) flow experiments were conducted. The concentration of the NaOH, which reacts with the limiting reagent (acid), was varied. Several turbulent mixing-limited reactions models (Eddy Breakup Model, Eddy Contact Model and Multiple-Time-Scale Model) were compared with the reactive flow data (2,2 DMP conversion to acetone and methanol).

This work has revealed a number of gaps in the current modeling and experimental approaches. For example, the mixing-limited reagent is often diluted with a solvent and reacted with an excess co-reagent. These engineering controls are not accurately depicted by some of the mixing-limited reaction models. Other key issues include the resolution the turbulent flow field and the need for models to simulate the mixing and reaction of intermediates. This presentation will explain these gaps and invite the academic community to provide solutions.

534878 Validation of the Coalescence-Dispersion Model for Complex Chemical Reactions

Abstract:

ABSTRACT

"Validation of the Coalescence-Dispersion Model for Complex Chemical Reactions"

Gary K. Patterson, Professor Emeritus

Missouri University of Science and Technology

Chemical and Biological Engineering

Rolla, Missouri 65401

The integration of the coalescence-dispersion (c-d) model of mixing into a CFD simulation code has been extensively tested. The c-d model simulates mixing by assuming that the fluids may be represented by mobile fluid points which carry the mass, concentrations and other properties of the fluids. Upon coalescence of two randomly chosen fluid points, their properties are averaged (concentrations, enthalpies, etc.). Upon dispersion (separation) the two fluid points have identical properties. In a c-d simulation of a chemical reaction, the reactions occur for the length of the time increment between flow events as if the fluid points are batch reactors. Use of the c-d model avoids the need for a closure model for the mixing in the presence of chemical reactions. Such closures have been successful only for single reactions and for two competing reactions.

Integration of the c-d model into a CFD simulation requires that the mobile fluid points move, mix and disperse in a way dictated by the CFD simulation. Each volume segment of the CFD simulation is assigned a number of fluid points which move from segment to segment as dictated by the CFD simulation. This is accomplished by a subroutine linked to the CFD code that accesses the velocity and mixing rate predictions of the CFD. A critical aspect of the c-d model is in the most effective choices of the time increment and number of fluid points in each CFD volume segment. This is dependent on the number of c-d events during each time increment and flow volumes from segment to segment. Of course, the accuracy of the CFD simulation is critical for the accuracy of the c-d simulation.

Validations of c-d simulations have been done by comparing results with well-established experimental data for single and multiple reactions in various types and sizes of chemical reactors. Examples of validation data used are experiments by Paul and Treybal, Penney and coworkers, Middleton and coworkers, Bourne and coworkers and by the author and students. The validations show that the c-d simulation method can be a valuable addition to experimental research on multiple competitive reactions where there is interest in yields of chosen products.

537344 A Direct Numerical Simulation of a Marginally Turbulent Stirred VesselÂ Equipped withÂ a Rushton Turbine

Abstract:

New results will be presented from a Direct Numerical Simulation (DNS) of the just turbulent flow field, at Re=7300, in a stirred vessel provided with baffles and driven by a Rushton turbine. The simulations have been carried out by means of a Lattice Boltzmann solver, written in C, on a lattice comprising some 30 billion nodes, carrried out at the Irish Centre for High-End Computing ICHEC in Dublin.

The results of this simulation are compared with experimental data from the literature and with the results of the less refined DNSs reported earlier. The main variables of interest are the turbulent kinetic energy and its dissipation rate, phase averaged as well as phase resolved, which can be calculated from the outcome of such a DNS at any position in the vessel. Also turbulent spectra can be obtained in this way and some spectra will be presented for some positions in the vessel.

535761 Blending of Miscible Fluids

Abstract:

The blending process of binary miscible fluids is governed by both fluid advection and fluid diffusion. In low Reynolds number systems, where flow is slow and often one-dimensional, blending is diffusion-limited, and the effects of advection on blending are small. In high Reynolds number systems, where the fluids are fully turbulent, miscible fluid blending is governed by advection and the effects of diffusion are minimal. Between these two extremes, the blending behavior is a more complicated hybrid of both transport processes. This complexly is exacerbated in systems with strong difference between the two fluid densities and viscosity.

Traditional computational approaches for predicting the blending of miscible fluids are successful only in the limit of very low or very high Reynolds numbers. The limitation is motivated by the spurious effects of numerical diffusion, which tend to overstate blending. In this work, we show how DNS and LES simulations can be combined with high-resolution flux-limiting schemes to predict the blending of miscible fluids across a range of Reynolds and Schmidt numbers. After discussing strategies for controlling numerical diffusion, we apply this approach to model miscible fluid transport in both agitated systems and continuous flow systems.

538613 Deconstruction of Mixing Processes Using Computational Fluid Dynamics and Z-Transform

Abstract:

Computational fluid dynamics (CFD) is used in studying the performance of mixing processes to complement experimental work by making accessible to the investigator measurements that are difficult to obtain physically. Motivated by the desire to enhance the modeling toolset, we have in recent years demonstrated the use of CFD with z-transform, the discrete-time counterpart of the Laplace transform, to provide a powerful approach for simulating and analyzing the dynamics of mixing processes. Examples that we have illustrated include the open-loop response of a lumped-parameter mixing process, the open-loop response of a distributed-parameter mixing process, the closed-loop response of an unstable mixing process, and the general characterization of mixing processes [AIChE Annual Meeting Papers 2014-354913, 2015-404589, 2016-449915, and 2017-486416].

By the conventional approach, continuous-flow mixing processes are typically modeled as ideal continuous stirred-tank reactors (CSTR) or plug-flow reactors (PFR), or combinations thereof. Real mixing processes, however, do not conform perfectly to the CSTR and PFR paradigm. In order to optimize real mixing processes or to troubleshoot problematic ones, it is desirable to transcend beyond the CSTR and PFR blackbox approach and describe the dynamic behavior of the nonideal mixing processes more accurately. In this presentation we extend our previous work on combining CFD and z-transform and demonstrate how one can deconstruct nonideal mixing processes into networks of smaller and simpler mixing and transport processes. The result allows us to identify and construct practical mathematical models that capture localized phenomena such as compartmentalization and short-circuiting in nonideal mixing processes.

541565 25 by 25: Chemical Engineering in the Next 25 Years

Abstract:

In 2008, the centennial year for AIChE, we asked 25 chemical engineers for their visions of the profession from around the US and the world: Nine faculty, eight from industry, and eight post-docs and graduate students. We posed four questions related to each individual's industrial sector or area of research, asking for extrapolation, new impacts on existing sectors, new sectors, and an optional, more open-ended comment on the future of the profession as a whole.

In this 110th year of AIChE, we have revisited these questions with many of the same people. Ten years has seen many developments in our profession, in society, and in our thought-leaders' experiences. Here we report on their perspectives, comparing them to the 2008 visions.

541542 The Future of Chemical Engineering Itself

Abstract:

Chemical engineeringâ€™s future is bright because it is vital to so much of society and industry -- and its future is complicated by its rapidly expanding universe of activities. It must represent depth in core skills centered on the chemical sciences including biology and materials. At the same time, it must be transdisciplinary; it must exemplify the need and value of reaching across disciplines in borderless ways.

One way is to create and disseminate a compelling vision and strategy for chemical engineering, research, and education for the future. AIChE and the National Academies are working together to that end. In 1988, the National Academies published "Frontiers in Chemical Engineering: Research Needs and Opportunities.â€ Its committee of authors outlined a wide-ranging roadmap to turn promising research opportunities into reality, while calling for researchers to embrace new frontiers.

Since 1988, that has happened -- along with many other transformative changes like the Web, fracking, personalized medicine, and social media. A new nationwide effort, aided by international contributions, will address both the identity of chemical engineering and its future directions. Its goal is to ensure that the people in chemical engineering are primed to meet society's needs and rapid changes.

544708 Accelerating Innovation through Academic-Industrial Partnerships

Abstract:

In 2011, The Dow Chemical Company launched the University Partnership Initiative (UPI) to sponsor collaborative research at leading U.S. Universities. This 10-year commitment aims to address fundamental questions in industrial chemistry and chemical engineering with alignment to Dowâ€™s next-generation process technology and product development efforts. To date, this portfolio of projects has generated an impressive array of intellectual property, publications, and numerous examples of the collaborative developments implemented at Dow. This presentation will introduce UPI program and highlight several contributions from our collaborators and leaders in the chemical engineering profession.

540559 Maximizing Uptime, Efficiency, and Safety of Industrial Operations through Early Risk Detection

Abstract:

Achieving zero incidents while maximizing process uptime and efficiency is a key vision for operating companies. An important building block towards this vision is establishment of a proactive risk mitigation culture, supported by effective systems and a strong management. Studies have shown that most of the incidents and unexpected process failures can be avoided if the operating teams get timely information about developing risks and take preventive actions early on. While advances have been made in process monitoring and automation in the past decades, there still remains significant technological and behavioral gaps that prevent plant operations from proactively mitigating process risks.

A modern industrial plant monitors thousands of parameters, generating upwards of 50-100 million data points every day. Thanks to recent breakthroughs in machine learning and artificial intelligence approaches, today there are autonomous systems that can sift through this data and point out meaningful and timely insight. This can help operating teams ascertain process issues that are hidden in the data, long before process variables reach critical levels.

In this presentation, a new autonomous system, Dynamic Risk Analyzer™ (DRA), will be introduced that points out early indicators of process issues using proprietary machine learning. The risk indicators enable operations team to respond and make required process changes, days and sometimes weeks ahead of any alarm, avoiding late stage (expensive) fixes. To realize these objectives fully, a proactive management workflow must also be in place. The presentation will discuss new initiatives and workflows developed by plant operations to implement a proactive culture. Real-life case studies from plants using DRA technology will be presented on how this culture change resulted in increased uptime, efficiency and safety of plant operations.

531309 Gaussian Processes for Hybridizing Analytical & Data-Driven Decision-Making

Abstract:

Surrogate models are widely appreciated in process systems engineering [1]. The typical setting focuses on expensive-to-evaluate, possibly uncertain functions. Examples include: modular process simulators [2], integrated gasification combined cycle processes [3], a carbon capture absorber [4], and many other applications, e.g. [5-8]. Resources are typically limited, so effective decision making requires data-efficient learning.

The data science and statistical machine learning communities typically focus on models learned solely from observed data. But chemical engineering applications may also require explicit, parametric models, e.g. modeling known process constraints, operations constraints, and cost objectives [9]. So, work has integrated semi-algebraic functions with those learned from data [10] or developed semi-physical modeling techniques [11, 12].

This presentation surveys the state-of-the-art in hybridizing analytical and data-driven decision making [13]. We consider three probabilistic modeling applications to these hybrid situations:

Design of experiments for model discrimination [14]. We bridge the gap between classical, analytical methods [15] and Monte Carlo-based approaches [16]. Classical methods may have difficulty managing non-analytical model functions and data-driven Monte Carlo approaches come at a high computational cost. We replace the original, parametric models with probabilistic, non-parametric Gaussian process surrogates learned from model evaluations. The surrogates are flexible regression tools that extend classical analytical results to non-analytical models, while providing us with model prediction confidence bounds and avoiding the computational complexity of Monte-Carlo approaches.

Multi-objective optimization [17, 18]. We make novel extensions to Bayesian multi-objective optimization in the case of one analytical objective function and one black-box, i.e. simulation-based, objective function. The resulting method has been applied to a bone neotissue application [19] and a more general test suite.

Scheduling plant operations under uncertainty. For processes with equipment degradation, we use Gaussian processes to approximate large-scale, mixed-integer optimization problems.

We close by offering a broad outlook on applying probabilistic surrogate models to chemical engineering. Statistical machine learning has recently attracted significant interest in process systems engineering [20]. Here we show that state-of-the-art research in Gaussian processes [21, 22] and probabilistic modeling more generally [23] can have a big impact on chemical engineering.

References

[1] Bhosekar A, Ierapetritou M. Advances in surrogate based modeling, feasibility analysis and and optimization: A review. *Comput Chem Eng*. 108: 250-267, **2018**.

[2] Caballero JA, Grossmann IE. An algorithm for the use of surrogate models in modular flowsheet optimization. *AIChE J*. 54(10):2633-2650, **2008**.

[3] Lang Y, Zitney SE, Biegler LT. Optimization of IGCC processes with reduced order CFD models. *Comput Chem Eng*. 35(9):1705-1717, **2011**.

[4] Cozad A, Sahinidis NV, Miller DC. Learning surrogate models for simulation-based optim-ization. *AIChE J*. 60(6):2211-2227, **2017**.

[5] Boukouvala F, Ierapetritou M. Surrogate-based optimization of expensive flowsheet modeling for continuous pharmaceutical manufacturing. *J Pharm Innov*. 8: 131, **2013**.

[6] Soepyan FB, Cremaschi S, Sarica C, and others. Estimation of percentiles using the Kriging method for uncertainty propagation. *Comput Chem Eng*. 93:143-59, **2016**.

[7] Shokry A, Ardakani MH, Escudero G, Graells M, Espuña A. Dynamic Kriging-based fault detection and diagnosis approach for nonlinear noisy dynamic processes. *Comput Chem Eng*. 106:758-76, **2017**.

[8] Tran AP, Georgakis C. On the estimation of high-dimensional surrogate models of steady-state of plant-wide processes characteristics. *Comput Chem Eng*. DOI 10.1016/j.comp-chemeng.2018.02.014, **2018**.

[9] Boukouvala F, Hasan MF, Floudas CA. Global optimization of general constrained grey-box models: new method and its application to constrained PDEs for pressure swing adsorption. *J Glob Optim*. 67(1-2): 3-42, **2017**.

[10] Boukouvala F, Floudas CA. ARGONAUT: AlgoRithms for Global Optimization of coNstrAined grey-box compUTational problems. *Optim Lett*. 11(5): 895-913, **2017**.

[11] Pearson RK, Pottmann M. Gray-box identification of block-oriented nonlinear models. *J Process Contr*. 10(4):301-315, **2000**.

[12] Cozad A, Sahinidis NV, Miller DC. A combined first-principles and data-driven approach to model building. *Comput Chem Eng*. 73: 116-27, **2015**.

[13] Boukouvala F, Misener R, Floudas CA. Global optimization advances in mixed-integer nonlinear programming, MINLP, and constrained derivative-free optimization, CDFO. *Eur J Oper Res*. 252: 701-727, **2016**.

[14] Olofsson S, Deisenroth MP, Misener R. Design of Experiments for Model Discrimination Hybridising Analytical and Data-Driven Approaches. *arXiv:1802.04170*. **2018**.

[15] Buzzi-Ferraris G, Forzatti P, Emig G, Hofmann H. Sequential experimental design for model discrimination in the case of multiple responses. *Chem Eng Sci*. 39(1): 81-85, **1984**.

[16] Vanlier J, Tiemann CA, Hilbers PAJ, van Riel NAW. Optimal experiment design for model selection in biochemical networks. *BMC Syst Biol*, 8(20), **2014**.

[17] Olofsson S, Mehrian M, Geris L, Calandra R, Deisenroth MP, Misener R. Bayesian Multi-Objective Optimisation of Neotissue Growth in a Perfusion Bioreactor Set-Up. *Computer Aided Chemical Engineering*. 40: 2155-2160, **2017**.

[18] Beykal B, Boukouvala F, Floudas CA, Pistikopoulos EN. Optimal design of energy systems using constrained grey-box multi-objective optimization. *Comput Chem Eng*. DOI 10.1016/j.compchemeng.2018.02.017, **2018**.

[19] Mehrian M, Guyot Y, Papantoniou I, Olofsson S, Sonnaert M, Misener R, Geris L. Maximizing neotissue growth

kinetics in a perfusion bioreactor: An *in silico* strategy using model reduction and Bayesian optimization. *Biotechnol Bioeng*. 115(3):617-29, **2018**.

[20] Lee JH, Shin J, Realff MJ. Machine learning: Overview of the recent progresses and implications for the process systems engineering field. *Comput Chem Eng*. DOI 10.1016/j.compchemeng.2017.10.008, **2017**.

[21] Damianou A, Lawrence N. Deep Gaussian processes. In *Artificial Intelligence and Statistics*. 207-215, **2013**.

[22] Deisenroth MP, Ng JW. Distributed Gaussian processes. In *32nd International Conference on International Conference on Machine Learning*. 37: 1481-1490, **2015**.

[23] Ghahramani Z. Bayesian non-parametrics and the probabilistic approach to modelling. *Phil Trans R Soc A*. 371(1984):20110553, **2013**.

537217 Mixing and Turbulence in Gas-Liquid Systems: Bridging First Principle and Application Needs

Abstract:

Most industrial systems in the chemical, process and nuclear industry are multiphase in nature. Multiphase models in Computational Fluid Dynamics that account for mixing, turbulence and interfacial exchange behavior offer the potential to transform the design of more efficient chemical and process systems. The effectiveness of its application relies on the accuracy and predictive ability of the models, which must, at the same time be integrated into an effective computational tool and allow optimal flexibility and scalability.

A large effort at MIT has been devoted to the development of a second-generation of multiphase- closures. In the past different groups have proposed different balances of drag, lift, turbulent dispersion and wall lubrication forces to represent the same set of experiments, producing comparably incomplete agreement with the experimental data, on a very limited set of conditions.

In this paper we will present a new set of momentum closures, where the near wall effects are included as a regularization of the turbulent dispersion based on the analytical near wall void fraction distribution, and the lift inversion is accurately predicted from the instability of bubble motion. The new formulation is assessed on a very comprehensive set of conditions, spanning the whole bubbly flow regime over a wide range of Eötvös and Reynolds numbers and showing a very good agreement with experimental data.

521478 Effect of Interfacial Forces on Mixing and Dispersion of Bubbles in Pipe Flows

Abstract:

Modelling the polydispersity features, e.g. size distribution, of the disperse phase is one of the crucial aspects of multiphase mixing simulations. Knowledge of the properties of the disperse phase is essential for accurate estimation of the interfacial heat and mass transfer occurring in practical and industrial mixing applications. In this regard, the population balance model coupled with the computational fluid dynamics, namely CFD-PBM, is a powerful tool for studying the evolution of the properties of the disperse phase and its interactions with the continuous phase.

Concerning the CFD, adopting the Eulerian-Eulerian framework, called Two-fluid model, is currently the only viable choice if the simulation target is the industrial-scale equipment. However, the performance of the two-fluid model depends strongly on the relations required to close the averaged field equations. The closure relations relevant to the momentum balance equation are termed as interfacial forces including mainly drag, lift, turbulent dispersion and wall lubrication forces. However, these models were mostly developed through simplistic theoretical/experimental methods that limits their applicability. Among the mentioned forces, lift and wall lubrication forces have been provoking ongoing debate over their physical explanation and corresponding suggested models.

In this work, the performance of two recent combinations of models proposed for the lift and wall lubrication forces are compared with each other using mono-disperse approach in a bubbly air/water two-phase system. Moreover, the potential for improving the results through adjustment of the lift coefficient is demonstrated. In the second part, the evolution of the bubble size distribution (BSD) is evaluated by employing the PBM method coupled with the Two-fluid model using the best setting for the interfacial forces. Eventually the results are compared with experimental data available.

540225 CFD Modelling of Multi-Regime Multiphase Flows

Abstract:

In many CFD simulations today, we would specify the flow regime of the flow we are considering so that we can select the appropriate modelling method and physics models for the simulation. For examples, we would use the Volume of Fluid (VOF) model for free surface flows, the Lagrangian particle tracking model for droplet and particle flows and the Eulerian multiphase (EMP) model for bubbly flows. However, many real life flows have a combination of different flow regimes that occur at different locations and they often change with time. For example, a stratified gas-liquid flow along a horizontal pipe, see Fig 1. As the gas velocity is increased, waves start to form at the interface and grow into slug flow. At the front and back of the slugs, gas is entrained into liquid to form bubbles and liquid into the gas to form droplets. In this case, we have a combination of stratified flow (ideally modelled using VOF), droplet and bubbly flows (using EMP). The slugs can form at any location along the pipe at any time and once formed they travel at a high speed along the pipe, so the flow regime changes with time and position. We can find many flows in different industries that have multiple flow regimes, for example boiling flows in a vertical channel, see Fig 2.

For these multi-regime flows, single flow-regime models are obviously not sufficient. Also, it is usually not possible to pre-describe the flow regime according to position as illustrated by the slug example above. Ideally, we want our CFD model to determine the flow regime and apply the appropriate models as part of the calculation. The development of such a multi-regime multiphase flow model is the focus of this paper.

The generic nature of the Eulerian multiphase (EMP) model allows it to represent different flow regimes in one CFD model. What we need is to build into the EMP solution procedure a method that can recognize and determine flow regime and with that select and assemble the appropriate physics models (such as interphase forces, sharp interface treatments, etc.) into the solution matrix for each computational cell. An important step in developing the multi-regime model is the ability to recognize, determine and therefore treat large scale interfaces (LSI) between two fluids. In the LSI regions we can apply the "VOF" type of numerical treatment, bringing in surface tension force model, including surface damping of turbulence, etc. The LSI region also marks the transition between two different flow regimes on either sides of it. In gas-liquid flows, it would usually be bubbly flow on the liquid-dominated side and droplet flow on the gas-dominated side. For these bubbly and droplet flow areas, we can apply the standard models (such as lift, drag, turbulence dispersion and virtual mass forces) for these flows.

In this paper we present some key aspects of large scale interface detection using the newly developed Adaptive Interface Sharpening (ADIS) numerical scheme together with demonstration and validation examples. Industrial flow examples are also provided to illustrate to what level of details the current model can capture and in so doing pointing to areas where further R&D work are needed.

Fig 1 Gas-liquid flow in a horizontal pipe

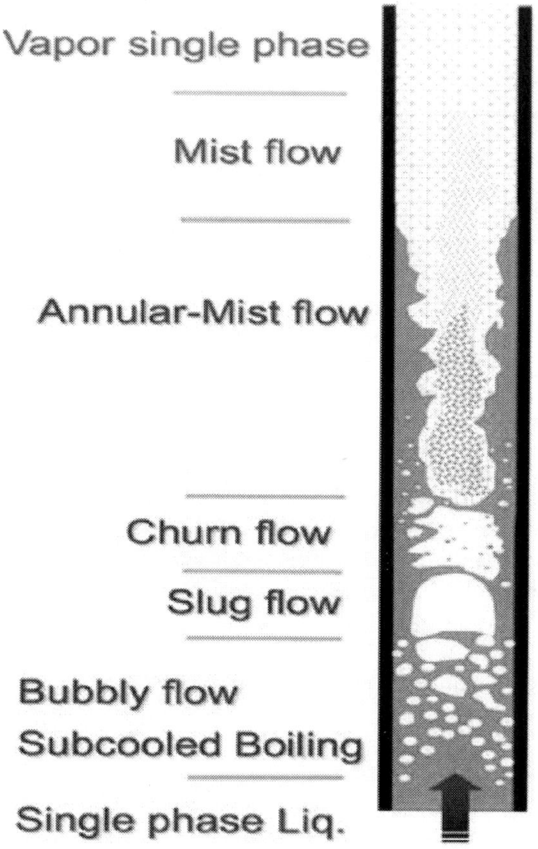

Fig 2 Boiling flow in a vertical channel

540875 Solid Suspension in Unbaffled Vessels Using Vertical Off-Center Agitators

Abstract:

Solids suspension via mechanical agitation has been studied extensively, with most of these studies being performed in baffled vessels. Baffling promotes axial flow that provides the desired performance with minimized capital and operating costs. While most studies focus on fully-baffled conditions, there are solids suspension studies that consider reduced baffling. An example is the work of Drewer *et al.* (1994) who reported advantages to using shorter baffles that do not extend to the tank base when agitating high concentration solid-liquid suspensions. Similarly, Myers *et al.* (2005) found that hydrofoil impellers can generate top-to-bottom motion with limited baffling, providing the opportunity to achieve solids suspension without full baffling. Baffling is not used in some applications, often motivated by the desire to promote cleanliness and avoid the potential for fouling. Very limited information is available concerning solids suspension in unbaffled vessels. In their study of solids suspension using angle-mounted agitators in unbaffled vessels, Myers *et al.* (2011) noted that the just-suspended speed in this configuration is typically seventy-five percent higher than with a vertically-mounted agitator in a fully-baffled vessel. Angle mounting is not always an acceptable approach, particularly for larger agitators whose weight leads to large forces on the agitator shaft and mounting structure. In these instances, vertical off-center agitator mounting may be used in an unbaffled vessel. Moving the agitator off the vessel centerline reduces swirl that leads to near solid-body rotation that often makes meeting agitation objectives challenging. Little information has been published about the performance of vertical off-center agitators in unbaffled vessels, and the study described in this presentation characterizes the effect of impeller type, size, and off-center location on the speed, torque, and power requirements for solids suspension in this configuration.

References

- Drewer, G. R., N. Ahmed, and G. J. Jameson, "Suspension of High Concentration Solids in Mechanically Stirred Vessels", *Proceedings of the Eighth European Conference on Mixing*, pages 41-48 (1994).
- Myers, K. J., E. E. Janz, and J. B. Fasano, "Baffling Approaches for Modern Axial-Flow Impellers", paper 335h, 2005 Annual AIChE Meeting.
- Myers, K. J., J. P. Herr, and E. E. Janz, "Solids Suspension with Angle-Mounted Agitators in Unbaffled Vessels", *Canadian Journal of Chemical Engineering*, Volume 89, pages 940-947 (2011).

535253 Predictive Modeling for Particle Dissolution in Mixing Tanks

Abstract:

Particle dissolution is an important process for a variety of applications in the food, pharmaceutics and the process industry. Accurate prediction of the dissolution time is essential to optimize process parameter like subsequent ingredient additions, RPM, and mixing vessel geometry such as baffles. In this study, an Eulerian-Lagrangian model was developed in STAR-CCM+ to simulate particle dissolution in mixing applications: benchtop beakers and pilot scale mix tanks. The continuous liquid phase and the Lagrangian particle phase was two-way coupled, therefore each particle dissolved based on its own local mass transfer coefficient and concentration gradient with continuous phase, thus capturing the spatio-temporal variation in the dissolution process. The model was validated against experiments for two different powders by comparing the dissolution time. The model showed excellent agreement in terms of predicting when the last particle would be dissolved to within seconds. Based on the simulation results, an analytical expression is suggested for the dissolution time to aid in design estimation.

527050 Hydrodynamics Characteristics of a Stirred Tank Provided with Angle-Mounted Impellers Using Computational and Experimental Approaches

Abstract:

Most industrial stirred tanks and reactors are provided with a centrally mounted impeller to ensure adequate mixing and with wall baffles to prevent the swirling effect that would otherwise occur in an unbaffled vessel. However, an alternative approach to reduce swirling and promote mixing in an unbaffled tank is to place the impeller asymmetrically and mounted on an angled shaft. This type of configuration is common in a number of industrial operations and especially in the pharmaceutical industry, since the absence of baffles reduces the potential for contamination and ensures thorough cleaning of the vessel between batches while introducing adequate top-to-bottom liquid recirculation of the vessel content and effectively eliminating swirling effects. Although industrially relevant and relatively common in the industrial practice, only limited information is available on the operation of tanks provided with angle-mounted impellers and even less is known of their hydrodynamics. Therefore, the objective of this work was to study the hydrodynamics of a scaled-down version (diameter: 316 mm) of a tall industrial vessel with an elliptical bottom and provided with two angle-mounted (by 5Â° off the vertical) A310 Lightnin impellers under different operating conditions in order to determine the mixing characteristics of the industrial system. Both Particle Image Velocimetry (PIV) and Computational Fluid Dynamics (CFD) were used to quantify the hydrodynamics and provide guidance on the optimal operation of these industrial system. The PIV apparatus was used to experimentally determine the velocity flow field in water at different agitation speeds so as to validate the CFD predictions. A multiple reference frame (MRF) computational approach and turbulence models based on the k-Îµ method were applied for all cases in the CFD simulations. The results revealed substantial agreement between the CFD predictions and the PIV experimental results. A complex flow in the vessel was observed. Due to the axial pumping action of the impellers, a downward flow impinging the vessel bottom was created resulting in the splitting of the flow. A portion of the zone near the bottom of the vessel, which is the preferential location for the sedimentation of settling solids in the liquid, was dominated a less well-mixed zone formation. Additionally, increasing agitation speeds had no significant effect on the presence of that zone.

530926 Experimental and CFD Study of Mixing Two Fluids of Different Properties and Its Application in Biological Drug Product Manufacture

Abstract:

During biological drug product manufacture (solution in syringe or vial), concentrated drug substance in solution, typically with high viscosity, is either used as is or diluted with buffer to desired concentration. The high viscosity drug substance is amenable for subcutaneous administration but is difficult to administrator through intravenous injection (IV) without dilution. While subcutaneous administration can be conducted at home by a non-healthcare professional, intravenous injection must be administrated by a healthcare professional. Therefore, highly concentrated drug substance with high viscosity is desired at the starting point of drug product process to allow flexibility between subcutaneous administration (used as is) and intravenous injection (used after dilution). Efficient dilution of highly concentrated drug substance can be achieved at an appropriate mixing speed in a given mixer. However, the max mixing speed is set by the sensitivity of the biological molecules to shear and interfacial stress. A speed too fast may raise stability concerns. Hence, an optimal mixing condition balances between efficiency and product stability.

In this study, sucrose solution of various concentrations is mixed with water or water like liquid to mimic the two-fluid mixing system in biological drug product formulation process(Sucrose solution: drug substance, water: buffer). The impacts of initial concentration of sucrose and impeller speed on the mixing time are investigated using both experiments and computational fluid dynamic (CFD) simulation at three different scale (impeller bottom driven) mixing tanks, 250ml, 10L and 400L respectively. The experimental results in 250ml and 10L tank show that there exists a minimum mixing speed at which two fluids homogenous mixing occurs. It is also found that the impact of the initial concentration of sucrose solution on mixing time decreases at the mixing speed increases. To scale up two fluids mixing system using the traditional dimensionless number like Reynold number or tip speed is challenging due to the complexity of two-fluid mixing. A CFD two-fluid mixing model recognizing the difference of viscosity and density between the two fluids is developed for scale up. The model captures the characteristics of two-fluid mixing as described in experiments. After validated by the experimental at 250ml and 10L tanks, the model is applied at a 400L mixing tank. The model suggests key mixing operation parameters based on the liquid volume in the tank. The suggested parameters work as expected from the model.

* Presenter.

533597 Mini Continuous Stirred Tank Reactors (mini-CSTR) for Cell and Tissue Culture Applications

Abstract:

An ideal cell culture system will provide a well-controlled, homogeneous, and steady environment for cells and tissues. For instance, well-controlled steady states would greatly benefit organ-on-chip experiments, stem cell culture, and tissue propagation (among other relevant biomedical applications). However, at present, no continuously stirred mini-reactors are commercially available for lab-scale culture applications.

We are developing simple, low-cost, and user-friendly miniaturized continuously stirred tank reactors (CSTRs) for biomedical and biotechnological agitations. These well-mixed mini-CSTRs will enable cost-efficient continuous culture at small scales. We use high resolution 3D-printing to fabricate these CSTRs and an Arduino platform to measure and control key parameters, such as temperature and pressure, in small portable incubators. Nutrients are fed by syringe pumps, and well-controlled low-speed (benign) agitation is provided by a custom-made magnetic system. Since the reactor behaves as a well-mixed reservoir, all bulk-liquid concentrations can be measured at the outlet stream, thereby greatly reducing the need for intrusive instrumentation. Here we present an exhaustive validation of the mixing performance of this system, including results from computational fluid dynamic experiments (CFD) at different agitation speeds. In addition, we present preliminary results of the use of this culture platform in two model applications: (a) the extended culture of breast cancer spheroids, and (b) the culture of Chinese Hamster Ovary Cells (the warhorse for biopharmaceutical production) for continuous production of biopharmaceutical compounds.

533915 Power Dissipation and Power Numbers for a Retreat-Blade Impeller in Pharmaceutical Mixing Tanks and Reactors Using an Experimentally Validated Computational Approach

Abstract:

Mixing in stirred vessels is a very common operation in many pharmaceutical processes. Glass-lined tanks and reactors are especially commonly used in the in pharmaceutical industry for the synthesis of Active Pharmaceutical Ingredients (APIs) because of their corrosion resistance, cleanability, and reduced contamination levels. In such cases, a retreat-blade impeller with a low impeller clearance off the tank bottom is commonly used in and the reactors are cylindrical and provided with a torispherical vessel bottom and equipped with a single baffle and a retreat-blade impeller (RBI). The power, *P*, dissipated by the impeller is critical to any mixing process since mechanical energy dissipation is needed to homogenize the vessel content, disperse immiscible phases, suspend solids, increase mass transfer, and, in general, produce the desired mixing level. *P* depends not only on the type of impeller used, agitation speed, and physical properties of fluid but also on the geometry of the system, including the location of the impeller in the mixing vessel. Despite their common use, little information is available on the hydrodynamics of RBIâ€™s is such systems and their mixing performance, including their power dissipation. In this work the power, *P*, dissipated by a RBI, and the non-dimensional Power Number, Po, in different fluids were obtained both computationally via Computational Fluid Dynamics (CFD) and experimentally, using a strain-gage based torque measuring device, for different Reynoldâ€™s Number values in a 61-Liter, scaled-down version of a typical glass-lined reactor for the synthesis of pharmaceutical APIs. Additionally, P and Po were obtained for a fully baffled system (four rectangular baffles) and for an unbaffled system. *P* and Po were obtained for a very large range of the Reynolds Number, Re ($1<Re<1,000,000$). Po was found to vary significantly with baffling type and Re. Excellent agreement was obtained between the CFD-predicted and the experimentally obtained values for Po. Correlation equations relating Po with Re using baffling type as the parameter were additionally obtained. These equations can be used by industrial practitioners to obtain Po and hence P for their system and thus help quantify and optimize pharmaceutical mixing processes typically involving the synthesis of APIs in this type of reactors.

AUTHOR INDEX